Water and health in Europe

A joint report from the European Environment Agency and the WHO Regional Office for Europe

WHO Library Cataloguing in Publication Data

Water and health in Europe : a joint report from the European Environment Agency and the WHO Regional Office for Europe / edited by Jamie Bartram...[et al.]

(WHO regional publications. European series ; No. 93)

1.Water supply 2.Potable water 3.Water quality 4.Water purification 5.Water microbiology 6.Water pollution, Chemical 7.Legislation 8.Europe I.Series II.Bartram, Jamie III.Thyssen, Niels IV.Gowers, Alison V.Pond, Kathy VI.Lack, Tim

ISBN 92 890 1360 5 (NLM Classification: WA 675)
ISSN 0378-2255

Text editing: David Breuer/Frank Theakston
Cover design: Sven Lund
Cover illustration: Patricija Kavaliauskaite, 5 years old, Kindergarten "Zilvitis", Jonava, Lithuania. This school participates in the European Network of Health Promoting Schools, a joint project of the European Commission, the Council of Europe and the WHO Regional Office for Europe.

Water and health in Europe

A joint report from the European Environment Agency and the WHO Regional Office for Europe

Edited by
Jamie Bartram, Niels Thyssen, Alison Gowers,
Kathy Pond and Tim Lack

WHO Regional Publications
European Series, No. 93

ISBN 92 890 1360 5
ISSN 0378-2255

PRINTED IN FINLAND

Contents

Foreword

Profound pressures on Europe's water resources affect health, the economy and sustainable development. Industrialization, intensification of agriculture, growing populations and increases in recreational demands accentuate the necessity for sufficient high-quality water resources. Conflicts between uses and users, coupled with the occurrence of natural disasters such as droughts and floods, highlight the need for sustainable management of water. Universal access to safe drinking-water and sanitation that protect human health and the environment is of primary concern in the pursuit of health and development. Nevertheless, water-related diseases occur throughout Europe, to which rural populations, socially excluded people and populations in areas affected by armed hostilities are especially vulnerable.

Although some aspects of water quality and supply have improved in some countries over the last decade, progress has been variable. Renewed emphasis is being placed on microbial quality and the acknowledgement of previously unrecognized and re-emerging microbial and other hazards. Many of the suggested solutions are as applicable today as they were a decade ago. However, major changes in administrative arrangements affected many countries in Europe in the 1990s, including the supply of water and sanitation services, land-use activities, pollution control and activities related to public health surveillance.

WHO's current health for all policy framework for the European Region is based on solidarity and a multisectoral approach to health, stating that "population exposure to physical, microbial and chemical contaminants in water, air, waste and soil that are hazardous to health should be substantially reduced, according to the timetable and reduction rates stated in national environment and health action plans". But health is also a human right, which presupposes that the prerequisites for health – of which sufficient quantities of good-quality drinking-water is one – are also a human right.

The European Union's fifth programme of policy and action in relation to the environment and sustainable development, from 1993, set targets for groundwater protection up to 2000. These include preventing permanent excess abstraction and all pollution by point sources and reducing diffuse source pollution. Water resources across Europe are shared and connected across national boundaries, and international cooperation therefore needs to be promoted to sustain Europe's water resources and to provide safe water for its inhabitants.

Partnerships and action were key themes of the Third Ministerial Conference on Environment and Health, held in London in June 1999. To this end the WHO Regional Office for Europe, in partnership with the United Nations Economic Commission for Europe, prepared a new Protocol on Water and Health to the 1992 Convention on the Protection and Use of Transboundary Watercourses and International Lakes. The Protocol was approved at the Conference and requires its signatories to take account of human health, water resources and sustainable development. This publication provides information on many of the issues covered by the Protocol, such as adequate supplies of drinking-water and sanitation, water for irrigation and recreational use, monitoring of hazards, and public participation in decision-making. The evidence presented was collected through an extensive coordinated data-gathering process, in which many organizations and individuals throughout the European Region have cooperated.

This publication takes forward some of the issues raised in Environment and health 1. Overview and main European issues*, jointly published by the Regional Office and the European Environment Agency, which highlighted the importance of the quality and availability of water in improving health. It is aimed at a broad readership and is* *intended to present the key issues in a format that can be appreciated by policy-makers, professionals and the general public alike.*

The 21st century will present a number of challenges to the aquatic environment. A coordinated approach to data collection, processing and management in Europe to support decision-making and to improve the reliability of environmental information will be essential to meet these challenges. I look forward to continuing successful cooperation between the Regional Office and the European Environment Agency.

Marc Danzon
WHO Regional Director
for Europe

Domingo Jiménez-Beltrán
Executive Director
European Environment Agency

Executive summary

Shortage of water may be the most urgent health problem currently facing some European countries, exacerbated by geography, geology and hydrology. In addition, climate change is predicted to have an influence, especially in coastal areas where flooding may disrupt sanitation infrastructure and thereby contaminate watercourses. Although many parts of Europe are currently well provided with fresh water, the water resources are unevenly distributed between and within countries, leading to shortages in many areas. The countries that are heavily populated and receive only moderate rainfall are particularly affected. Groundwater and surface water have a limited capacity for renewal, and pressures from agriculture, industry and domestic users affect the quantity of water resources. Both water quality and availability must therefore be integrated in long-term planning and policy implications concerning water management.

The extent of provision of piped drinking-water supplies to households varies across Europe and between urban and rural populations, with rural populations in the eastern part of the WHO European Region least well provided. Continuity of supply is also a problem in some areas. Inefficient use of water resulting from factors such as network leakage and inappropriate irrigation appears to be a significant problem.

The utilization of water for irrigation and for industry exerts pressure on water resources, which vary widely between countries and regions. One of the biggest pressures is agriculture and changes in irrigation practices. Agriculture accounts for approximately 30% of total water abstraction and about 55% of consumptive water use in Europe.

Population distribution and density are key factors influencing the quantity of water resources, through increased local demand for water in areas of high population density or limited precipitation.

Although high standards have been reached in some countries, outbreaks of waterborne diseases continue to occur across Europe, and minor supply problems are encountered in all countries. The immediate area of public health concern is microbial contamination, which can affect large numbers of people. The standard of treatment and disinfection of drinking-water is inconsistent across Europe and, especially where economic and political changes have led to infrastructural deterioration, can be insufficient. It appears that an increased number of outbreaks of waterborne diseases have occurred in countries and areas that have experienced recent breakdowns of infrastructure, resulting in discontinuous supply. Nevertheless, reliable data are lacking on the quality of the source water and the drinking-water supplied, and the detection and investigation of outbreaks are generally poor in most countries.

Inadequate sewerage systems are a significant threat to public health. A number of countries identify private and small public supplies as those most liable to receive insufficient treatment or to have insufficient protection for groundwater sources, and thus to be of poor quality. Poor infrastructure may be associated with financial constraints and/or organizational disruption. Nevertheless, the installation of advanced treatment works in large supplies is increasing in many countries, although occasional outbreaks of waterborne diseases are reported even in countries with high standards of supply. No clear trends are detectable, however, and international comparability of data is poor, hindering the development of regional assessments and evaluation of progress.

Numerous chemicals are found throughout the aquatic environment, but evidence of any effect on human health, except for effects arising from accidental releases, is often difficult to obtain. Problems of significant chemical contamination are often localized and may be influenced by geology or anthropogenic contamination. Concern about the effect of agriculture on the quality of water resources is often related to diffuse sources – contamination by agricultural chemicals, nutrients and microbial pathogens in particular.

Eutrophication is a major threat to European surface waters. Common fertilizers contain varying proportions of nitrogen, phosphorus and potassium. The use of fertilizers varies between countries, depending on the economic situation and predominant agricultural practices. Although such point sources of pollution as sewage discharges may contribute significantly to nutrient enrichment in some regions, diffuse sources – particularly agriculture – are the major contributors. In some countries, the proportion of water pollution caused by diffuse sources is steadily increasing.

Industrial demand and effects on water quality may be especially pertinent to urban areas with high populations, as industry is traditionally located in these areas. The amount of water used by industry and the proportion of total abstraction accounted for by industry vary greatly between countries. Abstraction for industrial purposes in Europe seems to have been decreasing since 1980. Industrial processes produce contaminated wastewater that may be released into marine and fresh surface waters, either directly or following treatment. Contamination may persist for several decades.

Considerable evidence has accrued linking the quality of bathing water with minor illnesses. The use of water for recreational activities is intrinsically linked to economics through the tourism industry, and the quality of such water is thus of considerable importance to tourism-dependent communities.

Although some improvements have been made over the past decade, coordinated efforts are still needed to ensure that Europe's population is supplied with wholesome and clean drinking-water and has access to safe recreational water. These include measures to control demand and to prevent, contain and reduce contamination by improving water and sanitation management at the international, national and local levels. One particular problem that has been highlighted in compiling this publication is the need to harmonize monitoring procedures where possible. Incorporating education and awareness initiatives is pivotal to the success of improved and harmonized monitoring programmes and to ensuring the safe use of water.

Additional efforts are required to sustain the European Region's water resources and to provide safe water for its inhabitants, both for drinking and for other purposes. Irrigation, drinking-water supply, industry, agriculture and leisure make competing demands on the quality and quantity of these resources, in addition to the need for water to maintain the aquatic ecosystem *per se.* Management of water has become fragmented because of the existence of diverse stakeholders and regulatory perspectives. Pollution control measures have traditionally targeted point rather than diffuse sources of pollution.

Trends in water management in Europe include moves towards catchment-level management, improved intersectoral coordination and cooperation, and frameworks facilitating stakeholder participation. This approach is developed by the European Union in its Water Framework Directive, which sets targets for good ecological status for all types of surface water bodies and good quantitative status for groundwater.

The roles of government and especially the private sector in water management, and in drinking-water supply and sanitation in particular, are being radically reappraised. The extent of this varies across Europe. International action plans and conventions have been agreed on, with targets for reducing pollution and measures necessary to reach the targets.

Partnerships and cooperation are needed between the environment and health sectors at all levels of government to disseminate technology, to improve management and to provide financial and institutional support to ensure access to safe water and sanitation for all. Integrated management systems must be adopted to ensure that the conflicting uses are managed in an effective manner to ensure safe use. Not only should long-term management be considered, but responses are required to unexpected events such as natural disasters or accidents with large-scale effects that can heavily influence the quality and quantity of water used for consumption.

Experience suggests that international management agreements develop most rapidly when a body of water is shared or bordered by a small number of countries at a similar level of economic development. The Convention on the Protection and Use of Transboundary Waters and International Lakes provides a strong focus for future integrated management of water bodies.

This publication aims to integrate this information on the state of the raw water sources with information gathered on the quality and provision of potable water and the impact on human health. The state of water resources in Europe has been reviewed, considering both availability and quality. This book assesses the accessibility and quality of potable supply across the Region and describes the public health implications of inadequate and contaminated sources.

Data collection

Published sources and collections of data were used to collate information on geographical features (land area, area under agriculture, irrigated area), water resources (availability and use) and water and sewerage infrastructures. EEA National Focal Points were asked to validate the data collected on their countries and to fill in any gaps, where possible.

At the same time, a questionnaire in two parts was sent to WHO country contacts. The first part requested information relating to drinking-water supply, including the regulatory infrastructure and monitoring requirements. Detailed information on the concentrations of certain parameters (nitrate, fluoride, arsenic, pesticides, total coliforms, faecal coliforms and faecal streptococci) was requested. General comments on contaminants that posed particular problems and any problems affecting the quality or continuity of supply were also sought. The second part of the questionnaire related to waterborne diseases. Detailed information was requested on the recorded incidence of and deaths from specific diseases (methaemoglobinaemia, dental and skeletal fluorosis, crytosporidiosis, giardiasis, hepatitis A, cholera, typhoid fever, amoebic dysentery, bacterial dysentery, amoebic meningo-encephalitis, severe diarrhoea and gastroenteritis) and any outbreaks or cases linked to drinking-water. General comments on any major problems in preventing waterborne diseases were also requested.

In countries where there was only one contact (EEA or WHO), these were asked to address both the questionnaire and the data on environmental features. In addition, contacts were asked to submit any reports or other information that would be of use in researching the project, and many volunteered information from their own experience.

In addition to the data collected specifically for this book, existing data holdings and published sources and texts were widely used as the basis for the more general sections.

Acknowledgements

Representatives of all Member States of the WHO European Region were invited to complete a questionnaire on aspects of drinking-water quality and waterborne diseases and to provide national reports, where available. Replies were received from just over half the Member States. Information on aspects of water resources and infrastructure and general data relating to land use and population were also requested. Published literature and previous assessments by the European Environment Agency were used extensively in preparing this publication. The current members of the Agency are the 15 European Union countries plus Iceland, Liechtenstein and Norway, whereas the WHO European Region is much larger (see table below). Respondents to the questionnaire, together with other experts from all Member States of the European Region were invited to review the first draft; the contributions of the individuals concerned were central to completion of the work and are much appreciated.

Contributors of data

G. Deliu, National Environment Agency, Albania
B. Reme & M. Afezolli, Ministry of Health and Environment, Albania
C. Vendrell Serra & M. Coll, Ministry of Health and Welfare, Andorra
J.-P. Klein, Federal Ministry of Labour, Health and Social Affairs, Austria
K. Kranner, Bundeskanzleramt, Austria

Member States (•) of the WHO European Region (WHO) and of the European Environment Agency (EEA)[a] and responses received to the questionnaire (x)

Country	WHO	EEA	Country	WHO	EEA
Albania	• x		Liechtenstein		• x
Andorra	• x	x	Lithuania	• x	x
Armenia	•		Luxembourg	• x	•
Austria	• x	• x	Malta	• x	
Azerbaijan	•		Monaco	• x	
Belarus	• x		Netherlands	• x	• x
Belgium	• x	•	Norway	• x	• x
Bosnia and Herzegovina	• x		Poland	•	
Bulgaria	•		Portugal	•	•
Croatia	• x		Republic of Moldova	• x	
Czech Republic	• x		Romania	• x	
Denmark	•	• x	Russian Federation	•	
Estonia	• x		San Marino	•	
Finland	• x	• x	Slovakia	• x	x
France	• x	•	Slovenia	• x	x
Georgia	•		Spain	• x	• x
Germany	• x	• x	Sweden	• x	• x
Greece	• x	•	Switzerland	• x	
Hungary	• x		Tajikistan	•	
Iceland	• x	•	The former Yugoslav Republic of Macedonia	•	
Ireland	• x	• x	Turkey	• x	
Israel	•		Turkmenistan	•	
Italy	•	•	Ukraine	•	
Kazakhstan	•		United Kingdom	• x	• x
Kyrgyzstan	•		Uzbekistan	•	
Latvia	• x	x	Yugoslavia	•	

[a] The European Environment Agency also collects data from countries other than its members (the European Union countries, Iceland, Liechtenstein and Norway), such as countries in central and eastern Europe.

F. Delloye, Direction générale des Ressources naturelles et de l'Environnement, Belgium
K. De Schrijver, Health Inspection, Belgium
D. D'Hont Lin, Aminal-Water, Belgium
Ministry for the Wallonia Region, Belgium
A. Teller, Cellule interrégionale de l'Environnement, Belgium
B. Ljubič, Minister of Health, Bosnia and Herzegovina
K. Capak, S. Sobot & B. Borcic, Croatian National Institute of Public Health, Croatia
R.N.B. Havlik, Ministry of Health, Czech Republic
T. Trei & S. Tarum, National Board for Health Protection, Estonia
P. Ahola, Ministry of Social Affairs and Health, Finland
O. Zacheus, National Public Health Institute, Finland
J.L. Godet & M. Isnard, Ministry of Health, France
B. Clark, H. Hoering & F. Schiller, Umweltbundesamt, Germany
M. Kramer, M. Exner & G. Quade, Rheinische Friedrich-Wilhelms-Universität, Germany
H.M. Römer, Bundesministerium für Gesundheit, Germany
S. Kitsou & V. Karaouli, Ministry of Health and Welfare, Greece
M. Csanady, National Institute of Public Health, Hungary
G. Steinn Jonsson, Environmental and Food Agency, Iceland
T. Power, Department of Health, Ireland
L. Drozdova, National Environmental Health Centre, Latvia
Ministry of Welfare, Latvia
Amt für Lebensmittelkontrolle, Liechtenstein
I. Drulyte, National Nutrition Centre, Lithuania
R. Petkevicius, Institute of Hygiene, Centre of Environmental Medicine, Lithuania
A. Zegrebneviene, Communicable Diseases Prevention and Control Centre, Lithuania
P. Hau, Division of Sanitary Inspection, Directorate of Public Health, Luxembourg
L. Licari, Department of Health Policy and Planning, Malta
R. Fillon, Ministre d'Etat, Monaco
C. Gastaud, Direction de l'Action sanitaire et sociale, Monaco
Mr Viora, Service du Contrôle technique et de la Circulation, Monaco
J.F.M. Versteegh, National Institute of Public Health and the Environment, Netherlands
T. Krogh, C. F. Nordheim & V. Lund, National Institute of Public Health, Norway

P. Torgersen, Norwegian Board of Health, Norway
I. Abreu, Ministry of Health, Portugal
N. Opopol, Ministry of Health, Republic of Moldova
A. Dumitrescu & I. Iacob, Institute of Public Health, Romania
H. Zajicova, Water Research Institute, Slovakia
Ministry of Health, Slovakia
M. Macarol-Hiti & M. Jereb, Ministry of Health, Slovenia
O. Tello Anchuela, Istituto de Salud "Carlos III", Spain
P.A. Garcia González, Ministry of Health and Consumer Affairs, Spain
M. Eriksson & H. Wahren, National Board of Health and Welfare, Sweden
B. de Jong, Swedish Institute for Infectious Diseases, Sweden
M. Wiman, Swedish Environmental Protection Agency, Sweden
P. Grolimund, Swiss Agency for the Environment, Forests and Landscape, Switzerland
P. Studer, Swiss Federal Office of Public Health, Switzerland
E. Mitchell, Department of Health and Social Services, Northern Ireland, United Kingdom
R. Scott, Drinking Water Inspectorate, Northern Ireland, United Kingdom
T. Hooton, Agriculture, Environment and Fisheries Department, Scottish Executive, Scotland, United Kingdom
S. Pescold, Water Services Unit, Scotland, United Kingdom
K. Andrews, S. Nixon & H. Horth, Water Research Centre, United Kingdom
F. Pollitt, Department of Health, United Kingdom
O. Hydes & J. Hilton, Drinking Water Inspectorate, United Kingdom
M. Rutter, Public Health Laboratory Service, United Kingdom

Reviewers

The editors are grateful to the following organizations and individuals for their comments.

B. Reme, Ministry of Health, Albania
J. Grath, Austrian Working Group on Water
Ministry for the Wallonia Region, Water Division, Belgium
K. Capak, Croatian National Institute of Public Health, Croatia
J.K. Fuksa, TGM Water Research Institute, Prague, Czech Republic

L. Romanovská, Ministry of Health, Czech Republic
P. Jantzen, National Board of Health, Denmark
B. Norup, Denmark
T. Säynätkäri, European Environment Agency National Focal Point, Finland
O. Zacheus, Finland
M. Exener, Hygiene-Institut der Universität Bonn
H. Horing, Umweltbundesamt, Institut für Wasser-, Boden- und Lufthygiene, Germany
A. Pintér, Hungary
S. Velina, Department of Public Health, Ministry of Welfare, Latvia
P. Bockmuehl, European Environment Agency National Focal Point, Liechtenstein
I. Drulyte, Lithuania
B. Kroes, European Science Foundation, Netherlands
Z. Kamienski, State Inspectorate for Environmental Protection, Poland
I. Iacob, Water Hygiene Unit, Institute of Public Health, Romania
J. Jezny, European Environment Agency National Focal Point, Slovakia
M. Kollárová, Ministry of Health, Slovakia
B. Metin, Ministry of Health, Turkey
G. Rees, Robens Centre for Public and Environmental Health, University of Surrey, United Kingdom
J. Cotruvo, NSF International, USA
R. Enderlein, United Nations Economic Commission for Europe
R. Bertollini, B. Menne & G. Klein, WHO Regional Office for Europe
P. Bourdeau, Chairman, Scientific Committee, European Environment Agency
R.A. Breach, European Union of National Associations of Water Suppliers and Waste Water Services (EUREAU), Commission 1

Thanks are also due to Concha Lallana, Claudia Koreimann, Helena Horth and Kevin Andrews, Water Research Centre, United Kingdom; and to Francesco Mitis, Enrico Nasi, Nicoletta di Tanno, Grazia Motturi and Daniella Boehm, WHO European Centre for Environment and Health, Rome Division, Italy.

1

Introduction

WATER – THE BASIS FOR DEVELOPMENT AND WEALTH

Over the centuries, proper management of the vital resource of water has led to developments and improvements in health across the European Region. Effectively managed water-supply and resource-protection systems generate the indispensable basis for agricultural and industrial production. Throughout the Region, urban and rural development have thrived where water sources have been effectively managed. In many growing cities in the Region this process started as early as the 15th and 16th centuries, but at least five decades of the 19th century saw water as a central preoccupation of many state and industrial leaders. As a result, in the first half of the 20th century, life expectancy increased, food became more healthy, infant mortality decreased and a number of major diseases no longer posed a serious threat to health. Scientific and technical development has led to excellent water supplies for household, farming and industrial purposes all over the Region. With a few exceptions, by the mid-1970s the Region as a whole was on the way to eradicating water-related diseases and to guaranteeing safe water for all.

Industrial development and wealth have depended on a safe, reliable and well managed water supply. It has been demonstrated to be the single most effective investment in economic and social development, and no other part of socioeconomic development has continued to be as incredibly cost-effective in relation to the wealth created. Over a wide range of income distributions, rich and poor countries alike have to invest less than 1% of the average income to ensure excellent water supply and resource management. This may be perceived as a major success, but it may also be the reason why the central role of water in societal development and wellbeing has lost focus. Always present and cheap, water has been taken for granted, and efforts to build and maintain both technical and human resources have lost their visibility and political weight. Instead of sustainable development, the responsibility of individuals, industry and civil servants has been progressively eroded.

Many countries have tried to loosen their ties to public water services, which were often seen as ineffective, bureaucratic and expensive. The new approach should release public budgets by transferring various responsibilities to private business. In so doing there is a certain risk that only the financially attractive portions of the supply systems will be "bought" by private companies. In some instances this may lead to well managed private water treatment facilities depending on ineffectively protected resources and badly maintained distribution systems. Once these responsibilities have been split and public services have been downsized it is much more difficult to re-establish a holistic approach to water management from the source to the customer.

In the western part of the Region, countries were entering this experiment from a comfortable situation, with improving resource protection, a well established legal framework and an educated democratic society. Those in the east, on the other hand, were confronted with complete political reorganization, with pressure to downsize their public services and at the same time to open their countries to strong market forces; this was not always to the benefit of the citizens, and in many countries there was no chance of keeping governmental control over the social and public health consequences of rapid investment.

Combating the creeping crisis in fresh water – competition or partnership?

As populations have increased and economies have grown, the competition among agriculture, industry and urban areas for limited water supplies has intensified. Modern forestry practices, the intensification of agriculture and the diversion of water for irrigation place additional stress on water resources. Widespread mismanagement of water resources in the past among all sectors – industry, farming, large urban populations and small communities alike – has contributed to a growing crisis in the management of fresh water resources. The worsening local conditions in the privatized water supply in the United Kingdom have been scarcely recognized, whereas water abuse in cotton farming under the Soviet regime was publicly known. The continuing water crisis in farming regions in southern Europe remains largely unchanged, even though considerable quantities of water diverted or pumped for irrigation are wasted. The abuse of large water systems such as the Syr Darya and Amu Darya Rivers (which feed 60 billion cubic metres annually into the Aral Sea basin), the irreversible deterioration in surface water quality by urban and industrial waste, saline intrusion of coastal aquifers, and contamination of groundwater by nitrates are all examples of avoidable stress on water resources.

Public or private?

Inconsistent legislation, ineffective implementation of existing laws, loss of responsibility and staff in public supervisory agencies, and the weakening or destruction of the sanitary–epidemiological institutions in the eastern part of the Region have accelerated the destruction of resources. At the same time, the collapse of industry in eastern Europe relieved rivers, lakes and groundwater of some of the continuous discharge of pollutants. Incoherent and inconsistent European Union (EU) directives and disparate national policies have created a huge administrative burden, yet have only reduced the speed at which water resources are being destroyed. Despite a broad range of political action, the second assessment of Europe's environment conducted by the European Environment Agency *(1)* was unable to identify substantial improvements in water quality in Europe.

Technical requirements demanded by legislation have lagged far behind many inexpensive and effective means of improving water quality based on local awareness and interest. Examples such as resolving the ecological crisis in the Ruhr (Germany) between 1910 and 1960, to implementation of advanced wastewater treatment in Sweden and Switzerland in the 1970s and 1980s, and cooperation between water suppliers and farmers in several areas across the Region have given rise to the hope that partnership will prove cheaper and more effective than competition.

Price or value – the precious public good

Economic stability – rather than blind economic growth – is an important development objective in many countries. The financial burden on users to pay for water and sanitation is incredibly low compared to the health cost incurred by failure to provide safe water to everyone. In the overall context of increasing health care costs, water has to be highlighted as a central political issue. Over 30 million cases of water-related disease could be avoided annually through water and sanitation interventions. Investing in water supply and sanitation has produced benefits far greater than those directly related to the cost of treatment for water-related diseases.

The organizational structure of water services in the Region has seen an increasing trend towards the private sector. The services in England and Wales are wholly privatized, while France has a partly private system. A variety of other approaches can also be found, ranging from direct operation and management by local authorities to private enterprises governed separately by public administrations. The effectiveness of some of these privatization approaches still needs to be evaluated in terms of public health and economic criteria.

Services in central and eastern European countries are now predominantly operated by local administrations that have lost status and resources. International funding agencies find it difficult to invest in local sustainable water and sanitation services, partly because low (or no) prices for water cannot provide enough funds for the needed reconstruction of damaged networks and treatment facilities. Basic issues still need to be addressed, such as the provision of a continuous supply of water of adequate microbial quality.

Public valuation and participation

Consumers are extremely willing to pay for water of good quality. Consumption of bottled water is growing in a number of countries.

Many people lack confidence in the quality of their tap water and have therefore invested in household filtering devices, without knowing that most of these filters do not effectively control contaminants or pathogens. Public pressure and greater awareness have helped to create and carry out a number of pollution control programmes in recent decades in several European countries. Nevertheless, people are increasingly seeking cleaner water for recreation and are prepared to travel to find recreational water of good quality.

Participation by nongovernmental organizations and the business world is crucial to improving the management of water resources. Many nongovernmental organizations originate from local initiatives and are independent and self-managed. Their knowledge of local issues and conditions and their local contacts aid the motivation and awareness of local communities in advocating change. The cooperation of such organizations with public efforts at the community or national level is essential in providing an improved environment for human health. Public policies require the consent and sometimes the active participation of individuals. Accepting responsibility is an important and basic element.

Building new commitment

Having recognized the problem of conflicting requirements imposed by different pieces of legislation, the EU countries agreed to develop the Water Framework Directive *(2)* in order to produce an instrument for integrated water management, aiming to create a holistic framework for the protection of inland surface waters, transitional water (shallow water), coastal waters and groundwater. At the same time, the obvious need for action also outside the EU resulted in the decision by the European Member States of WHO to embark on developing a legally binding pan-European instrument. Focusing on health and wellbeing targets in all Member States, the Protocol on Water and Health *(3)* will foster partnerships in order to improve the outcome of water supply and resource management. Intersectoral action has been placed at the centre of the HEALTH21 policy framework adopted by the Member States of the European Region in 1998 *(4)*.

This type of action requires a well established and widely agreed database – not only about health concerns, but much more about ways of developing sustainable, healthy and economically sound water management systems. This publication gives an overview from many perspectives:

- comparison between countries of the availability of water resources;
- variation in data collection and density in various parts of the Region;
- different types of stress on water quality and quantity; and
- differences in economic valuation and pricing of water use and services.

The compilation of country reports hides a most important aspect of water management and health in Europe: almost every country includes a large variety of good and poor approaches to solving or creating problems related to water and health. Further work is needed to develop the general essence of many case studies of environmental excellence.

This publication can be a starting point for identifying excellent approaches. Most of these have been proven to be not only good for health and wellbeing but also economically sound and sustainable. There is no reason to implement "average" performance mechanisms when there are good opportunities to develop awareness and commitment for the best solutions. The motivation and expectations of individuals is high and must be encouraged and realized. The successful management of water resources depends on the ability and willingness of the regulators to meet those expectations.

2

European water resources

Distribution of resources

Geographical distribution

The water resources of a country are determined by a number of factors, including the amount of water received from precipitation, inflow and outflow in rivers and the amount lost by evaporation and transpiration (evaporation of water through plants). The potential for storage in aquifers and bodies of surface water is important in facilitating the exploitation of this resource by humans. These factors depend on geography, geology and climate.

Freshwater resources are continuously replenished by the natural processes of the hydrological cycle. Approximately 65% of precipitation falling on land returns to the atmosphere through evaporation and transpiration; the remainder, or runoff, recharges aquifers, streams and lakes as it flows to the sea.

Methods for calculating the availability of freshwater resources vary considerably from country to country, making comparison difficult. To overcome this, Rees et al. *(5)* have developed a method of estimating the renewable freshwater resources across the EU. This method uses data from hydrometric (river gauging) networks, supplemented by an empirical freshwater balance model that relates runoff to precipitation and potential evaporation. Freshwater resources vary considerably across the European Region: annual runoff ranges from over 3000 mm in western Norway, to 100 mm over large areas of eastern Europe, and to less than 25 mm in inland Spain (Fig. 2.1).

The average annual runoff for the member countries of the European Environment Agency (EEA) is estimated to be about 3100 km^3 per year (314 mm per year). This is equivalent to 4500 m^3 per capita per year for a population of 680 million *(6)*. The population of the WHO European Region is some 870 million, but figures for total runoff are not available.

Fig. 2.1. Long-term average annual runoff (expressed in mm) in the European Union and nearby areas

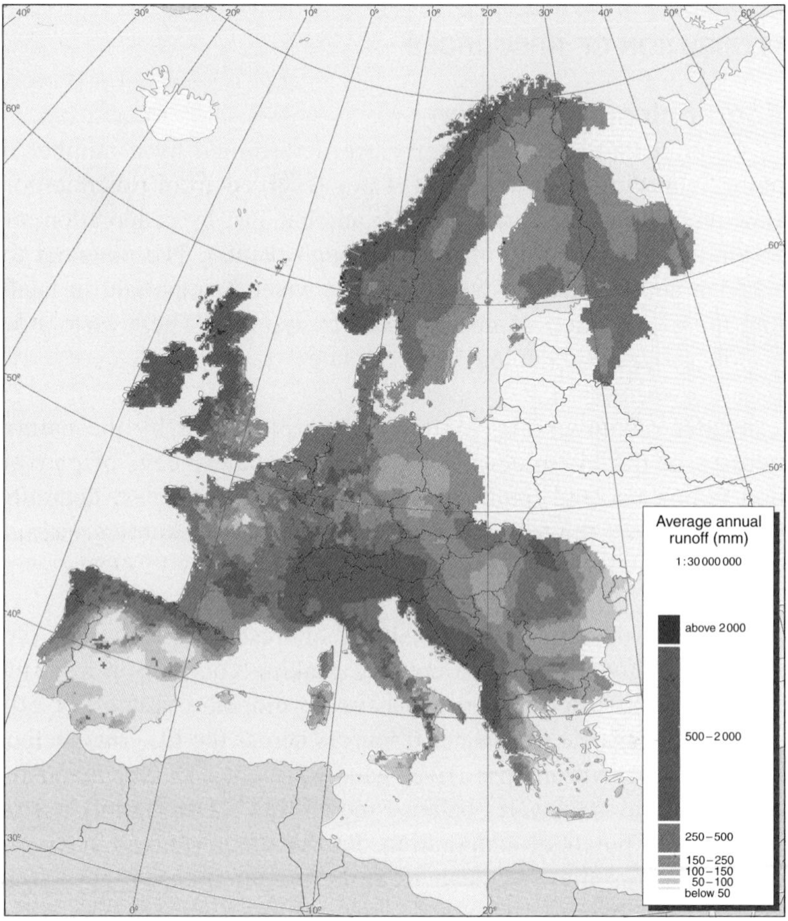

Source: European Environment Agency *(1)*.

Sustainable use of the freshwater resources can only be assured if the rate of use does not exceed the rate of renewal. The total abstraction of a country or area must not exceed the net water balance (precipitation plus inflow minus evaporation and transpiration). An excess of water abstraction over water use is especially prominent in the central Asian republics, the Russian Federation and Ukraine *(7)*. Achieving the correct balance between use and renewal requires reliable quantitative assessment of the water resources and a thorough understanding of the hydrological regime. Available resources must be managed carefully to ensure that abstraction to satisfy the various demands for water does not threaten the long-term availability of water. Sustainability also implies management to protect the quality of the water resources, which may include measures such as preventing contaminants from entering the water, and maintaining river flows so that any discharges are sufficiently diluted to prevent adverse effects on water quality and ecological status.

Population density also determines the availability of water per person. Population density varies widely across Europe, from fewer than 10 inhabitants per km^2 in Iceland, the Russian Federation and some of the central Asian republics (Kazakhstan and Turkmenistan) to over 300 per km^2 in the Benelux countries and San Marino and over 1000 per km^2 in Malta (Fig. 2.2).

On the continental scale, Europe appears to have abundant water resources. Latvia, for example, consumes only 1.3% of the natural renewable resources annually. However, these resources are unevenly distributed, both between and within countries *(8)*. Once population density is taken into account, the unevenness in the distribution of water resources per inhabitant is striking.

Many European countries have relatively little water available. Southern countries are particularly affected, with Malta having only 100 m^3 per capita per year (less than 5000 m^3 per capita is regarded as low; less than 1000 m^3 is extremely low and is commonly used as a benchmark of water scarcity; and above 20 000 m^3 per capita is considered high). Heavily populated countries with moderate rainfall in western Europe, such as Belgium, Denmark and the United Kingdom, are also affected, as are the Czech Republic and Poland in central Europe. Water resources are unevenly distributed and reported

Fig. 2.2. Population density in the WHO European Region

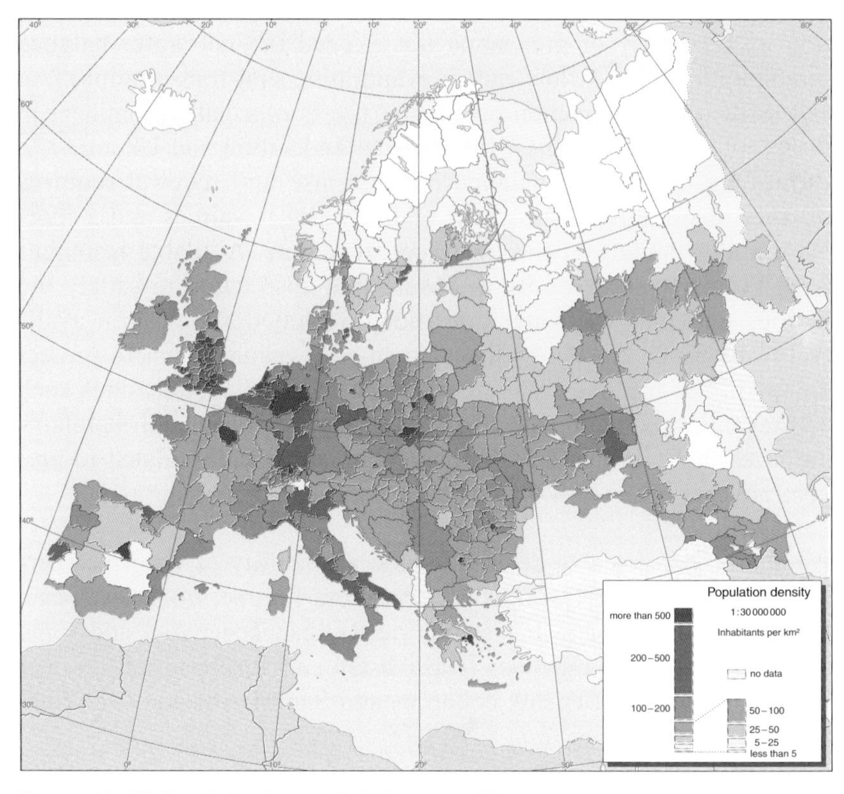

Source: Health for all database, WHO Regional Office for Europe.

to be declining in regions of the Russian Federation *(9)*. Twenty-seven per cent of the territory of the former USSR has insufficient water available *(10)*. Insufficient water resources are reported in southern Ukraine, the Republic of Moldova, the middle and lower reaches of the River Volga, the Caspian lowland, the southern parts of western Siberia, Kazakhstan and the Turkmenistan lowland. Only the Nordic countries with high rainfall that are sparsely populated have high water availability.

Precipitation varies between seasons and years. Countries or areas that usually have access to adequate water resources may suffer shortages at certain times of the year or in certain years.

Local demand for water in areas of high population density or limited precipitation may exceed the local availability of water. Excessive exploitation of groundwater sources in such cases not only threatens the future adequacy of the water supply but also affects the local environment: for example, the loss of wetlands, desertification, low river flow and, in the case of coastal aquifers, intrusion of salt water.

Most water used for all purposes in Europe is abstracted from surface water sources *(1)*. Groundwater comprises most of the remainder, with only a minor contribution from desalination of seawater, mainly in Italy, Malta and Spain.

There are some exceptions. Denmark abstracts 99% of its water from extensive groundwater reserves. In Latvia, groundwater and surface water are abstracted in approximately equal quantities. Where sufficient groundwater reserves are available, these are generally preferred as the source of public water supplies and, in many countries, provide the majority of drinking-water. Groundwater is generally of higher quality than surface water, and therefore requires less treatment before being suitable for public water supply. Protecting groundwater against contamination and excessive exploitation is therefore of great strategic importance.

Distribution in time – floods and droughts

Extreme hydrological events, such as floods and droughts, are a natural characteristic of hydrometeorological variability. Despite the progress in science and technology and increasing expenditure on drought amelioration and flood control, people are still vulnerable to extreme hydrological events, both in industrialized and in developing countries.

Floods

Floods can cause heavy damage. Seasonal fluctuations in water level and discharge, as well as inundation of riparian areas, are natural features of running water. However, the areas liable to flooding are often settled, and flooding interferes with human land use; damage can be enormous and many lives can be lost (Box 2.1).

Floods have been reported since ancient times, but during the 1990s in Europe and other parts of the world episodes of flooding, threatening

Box 2.1. The European flood of 1997

What happened?
In July 1997, Europe experienced some of the most disastrous flooding in its history. Vast areas of southern Poland, the eastern Czech Republic and western Slovakia were flooded after exceptionally heavy rain. At the worst-hit locations, as much water fell in a few days as usually falls in an entire year (for example, 585 mm in five days at one monitoring point in the Czech Republic). Many streams in the watersheds of the Oder, Elbe, Vistula and Morava Rivers overflowed their banks. The surges moved downstream, flooding communities and destroying houses and bridges. Industrial waste and sewage entered the floodwater, contaminating everything it touched: agricultural soil, stores, offices and homes.

The flooding affected one quarter of Poland, an area populated by 4.5 million people in nearly 1400 towns and villages. The towns of Opole, Klodzko and Wroclaw were devastated. In Poland, 400 000 hectares of agricultural land were affected, 50 000 homes destroyed, 5000 pigs and 1 million chickens lost, 170 000 telephone connections cut, 162 000 people evacuated and 55 people killed. Infrastructural damage included 480 bridges, 3177 km of road and 200 km of rail track. The total damage to Poland was estimated at US $4 billion.

In the Czech Republic, the flood caused US $2.1 billion in damage, 40 people were killed in the floodwater and 10 people died subsequently (of heart attacks and infections). About 2150 homes were destroyed and 18 500 damaged, and 26 500 people were evacuated. In Germany about 6500 people were evacuated. The costs in the most severely affected German *Land*, Brandenburg, were estimated to be US $361 million.

Underlying causes
The flooding was caused by extremely heavy rain, but the effect was intensified by human changes to the surroundings. In particular, the water retention potential of several of the flooded watersheds had been reduced by human intervention. Forests and riverine wetlands had been destroyed, mountain streams and rivers engineered, waterside vegetation destroyed, natural water-retention features removed and agricultural land drained, all of which reduced the absorptive capacity. Straightening and shortening of the Oder and Vistula had made them additionally susceptible to flooding.

Lessons learned
A change of attitude is required. Hazard prevention and response have to be seen as part of a dynamic interaction between people and nature. There must be more awareness and understanding of the interactions between human activities and natural systems.

human lives, property and infrastructure, seem to have increased. The effects are made worse by the expansion of human settlements and infrastructure to flood-prone areas. The trends have yet to be established unequivocally. Serious floods in Europe during the 1990s are shown in Table 2.1.

Table 2.1. Serious floods in Europe in the 1990s

River	Year	Fatalities	Damage costs	Remarks
Tazlau (Romania)	1998	107	€50 million	Breakdown of the Tazlau dam
Ouveze (France)	1992	Nearly 100	Not known	Camp site
Rhine/Meuse	1993/1994	10	€1100 million	
Po	1994	63	€10 000 million	Catchment area covered by up to 60 cm of mud
Rhine	1994/1995	None	€1600 million	Evacuation of 240 000 inhabitants in the Netherlands
Glomma and Trysil River Basins (Norway)	1995	None	€300 million	Largest flood since 1789
Pyrenean River	1996	85	Not evaluated	Camp site
Oder, Elbe, Vistula and Morava	1997	95	€5900 million	195 000 people evacuated; great material loss
Lena (Sakha, Russian Federation)	1998	15	1300 million roubles	51 295 people evacuated; complete disruption of transport system; great material loss

Source: European Environment Agency *(1)*.

The risk of flooding results from natural influences on the frequency of floods and human interventions in the hydrological cycle that influence the consequences of flooding.

Drought

The characterization of drought includes concepts that are not strictly meteorological or hydrological, such as social, economic or agricultural considerations.

A drought is an extreme hydrological event, whereas aridity is restricted to regions with low rainfall and is a permanent feature of climate. Recent European droughts have emphasized that the hazard is not limited to semi-arid countries and is a normal part of climate in all countries. Drought has a number of effects: loss of human lives (directly through thirst or indirectly through starvation or disease); loss of crops and animal stock; water supply problems, including shortages and deterioration of quality; increased pollution of freshwater ecosystems by concentration of pollutants; regional extinction of animal species by the absence of biotopes in drought periods; forest fires; wetland degradation; desertification; effects on aquifers; and other environmental consequences.

Climate change

Although regional differences are relatively high, most of Europe experienced increases in temperature of about 0.8 °C on average in the 20th century *(12–15)* and climate models predict global mean surface temperatures could rise by about 1–3.5 °C by 2100 *(16)*. Annual precipitation trends in the 20th century included enhanced precipitation in the northern half of Europe (north of the Alps to the Nordic countries), with increases ranging from 10% to close to 50%. The region stretching from the Mediterranean Sea through central Europe into the European part of the Russian Federation and Ukraine, by contrast, experienced decreases in precipitation by as much as 20% in some areas.

Predictions of climate change are subject to huge uncertainty. Even where the likely global trend appears to be clear, the response in individual regions may vary substantially from this. Thus, although global temperatures are predicted to increase by 1–3.5 °C by the year 2100 *(14)*, the actual rise in individual areas will differ significantly, and some regions may become cooler.

Similarly, global average precipitation is predicted to rise, but this increase is also likely to be regional. It is predicted that winter and spring precipitation will increase in Europe and summer precipitation will decrease, although the Mediterranean region and central and eastern Europe are expected to experience reduced precipitation *(17)*. The incidence of drought and heavy precipitation events is also therefore predicted to increase, which suggests implications not only for increased contamination resulting from run-off but also decreased groundwater recharge and an increased incidence of flooding.

Effects of climate change on the quantity of water resources

Complex interactions in time and space between precipitation, evaporation, discharge, storage in reservoirs, groundwater and soil make it difficult to model and analyse the influence of climate change on the hydrological cycle.

One of the basic mechanisms is that higher temperatures lead to higher potential evaporation and decreased discharge (which is also a function of precipitation, storage and topography). The storage in the soil serves as a buffer; in winter and spring, increasing precipitation normally generates higher discharges because the buffer is full and evaporation is low *(18)*. During the summer, storage is reduced by evaporation and transpiration, and the soil must be refilled before discharge begins. Changes in the hydrological cycle are much more variable than changes in other climatic factors. Seasonal to interannual variability in precipitation and temperature also accounts for some of the variability in hydrological characteristics in European river basins. Predictions about hydrology are difficult in Europe because anthropogenic factors, such as changes in land-use patterns and the drainage conditions of rivers and an increasing proportion of impermeable areas, strongly influence the European hydrological cycle *(18)*.

Predictions of the effect of climate change on river flow are uncertain, and the results of different models are highly variable. Arnell & Reynard *(19)*, for example, modelled river flows in the United Kingdom under various climate change scenarios and found that, under all scenarios, the concentration of flow was greater in winter. The models predicted that monthly flow would change by a greater percentage than annual flow, and different catchment areas would respond differently to the same scenario. The models indicated that progressive change would be small compared with variability over a short time scale, but that it would be noticeable on a decade-to-decade basis.

Cooper et al. *(20)* found that the effect of various climate change scenarios on aquifer recharge depended on the aquifer type, and that a scenario incorporating high evaporation produced the greatest change in hydrological regime.

The central emission scenario of the Intergovernmental Panel on Climate Change predicts a rise in sea level of 0.5 m by the year 2100. However, the predicted rise will not be uniform around the world. Polders, such as lowlands in the Netherlands and northern Germany, will be submerged. Catchment areas in flatlands depend on groundwater recharge, and changes in percolation can change the size of catchment areas *(21)*. On a local catchment scale, the distribution of water in the landscape can change even if annual discharge remains unchanged: whereas hilltops are severely stressed by droughts, areas with high groundwater levels may remain largely unaffected *(18)*.

Although the debate about changes in the frequency of floods is still open, an increase in rainfall during periods when soils are saturated (winter and spring) could increase the frequency and severity of floods. An increase in large-scale precipitation might lead to increased flood risks on large river basins in western Europe in winter *(18)*.

Hotter summers would lead to increased demand for water for irrigation purposes in already sensitive regions (such as the Mediterranean basin and central Asian republics), especially for soil with low storage capacity to handle summer water shortages *(18)*.

Water demand is likely to increase in some countries because of increasing irrigation, population growth or increased use of domestic appliances. If decreased water availability, because of climate change, is also considered, then an imbalance of supply and demand is likely. The potential influence of climate change on water resources (m^3 per capita per year) in some European countries is demonstrated in Table 2.2, although the variation between different climate scenarios should be noted.

Rivers

Surface water is vulnerable to contamination from many sources. Potential contaminants include agricultural chemicals and micro-organisms in run-off from agricultural land, chemicals in industrial discharges, and nutrients and pathogens from domestic sewage. Surface water is often

Table 2.2. Estimated water available (m^3 per person per year) in selected European countries in 1990 and 2050 based on projection of present climate conditions (change resulting from population growth and other non-climate-related factors) and three transient climate change scenarios

Country	Present climate, 1990	Present climate, 2050	Range of three climate scenarios, 2050
France	4110	3620	2510–2970
Poland	1470	1250	980–1860
Spain	3310	3090	1820–2200
Turkey	3070	1240	700–1910
Ukraine	4050	3480	2830–3990
United Kingdom	2650	2430	2190–2520

Source: McMichael et al. *(22)*.

used for many purposes other than water supplies (such as transport, irrigation and leisure), and this may affect the water quality. For example, fuel leaking from boats can potentially harm the aquatic life of the river and those who consume water or use it for recreation.

If a toxic contaminant adsorbs strongly to sediment its effect may not be immediately apparent, except on bottom-dwelling organisms, as the concentration in the water column will be reduced by the adsorption. However, contaminated sediments can act as a reservoir for the subsequent gradual release of chemicals into the water column. This may apply, for example, to historical releases of heavy metals by industry, organochlorine compounds such as polychlorinated biphenyls, and the sheep-dip chemical cypermethrin. Release of chemicals bound to sediment is especially likely when the sediment is disturbed.

The release of nutrients (especially nitrogen and phosphorus) that predominantly originate from agriculture and domestic sewage may result in eutrophication of vulnerable bodies of water. Such nutrient enrichment can cause significant changes in the balance of the aquatic ecology, often resulting in an algal bloom (see page 30).

In rivers, contamination from point sources is diluted and carried away from the source. However, multiple discharges along the course of a river can result in higher levels of contamination in downstream stretches.

Microbial contamination

Every effort should be made to achieve water quality that is as high as practicable. Protection of water supplies from contamination is the first line of defence. The microbial quality of surface water varies widely both temporally and spatially, and European countries report different trends without any consistent geographical pattern. Many rivers in Europe are significantly contaminated with microbes, arising from municipal wastewater and/or animal husbandry, that are of public health concern. In the bodies of water situated around the Aral Sea, especially in the Kzyl-Orda region of Kazakhstan and the Karakalpakstan Autonomous Republic in Uzbekistan, the high total microbe numbers present a risk of infectious waterborne diseases *(23)*. A comparative study (1981–1985 and 1986–1990) of the microbial pollution of the River Danube revealed that it is characterized by a high percentage of frequencies exceeding 100 000 coliform bacteria per litre. The high frequency of non-compliant drinking-water samples and presence of enterobacteriophages coincided with incidents of acute diarrhoea and viral hepatitis type A in the localities of Turnu-Magurele, Braila, Tulcea and Cernavoda *(24)*.

Waterborne sewage that is not exhaustively treated inevitably requires treatment for drinking and irrigation and may be unsuitable for recreational use. As far as possible, water sources must be protected from contamination by human and animal waste, which can contain a variety of viral, bacterial and protozoan pathogens and helminth parasites. Failure to provide suitable protection and adequate treatment will expose the community to the risk of outbreaks of intestinal and other infectious diseases.

Organic matter

The organic matter content of water is usually measured as the biochemical oxygen demand (BOD) and/or the chemical oxygen demand (COD). These terms are not directly comparable, but given the dual approach across Europe, general comparisons have to be attempted. In undisturbed rivers, typical BOD values are less than 2 mg oxygen per litre and those of COD 20 mg oxygen per litre.

In the rivers of Nordic countries, the organic matter content of anthropogenic origin is generally low. In many other countries of the European Region, BOD measurements exceeding 5 mg oxygen per litre (compared with less than 2 mg per litre in undisturbed rivers) have been recorded, especially in rivers subject to intense human and industrial use *(1)*. In a recent assessment, an average BOD exceeding 5 mg oxygen per litre, indicating significant organic pollution, was recorded at 11% of river stations throughout Europe. Nevertheless, 35% of all river stations had an average BOD of less than 2 mg per litre, indicating an acceptable organic matter content *(1)* (Fig. 2.3).

Fig. 2.3. Average annual concentration of organic matter in selected European rivers, 1994–1996

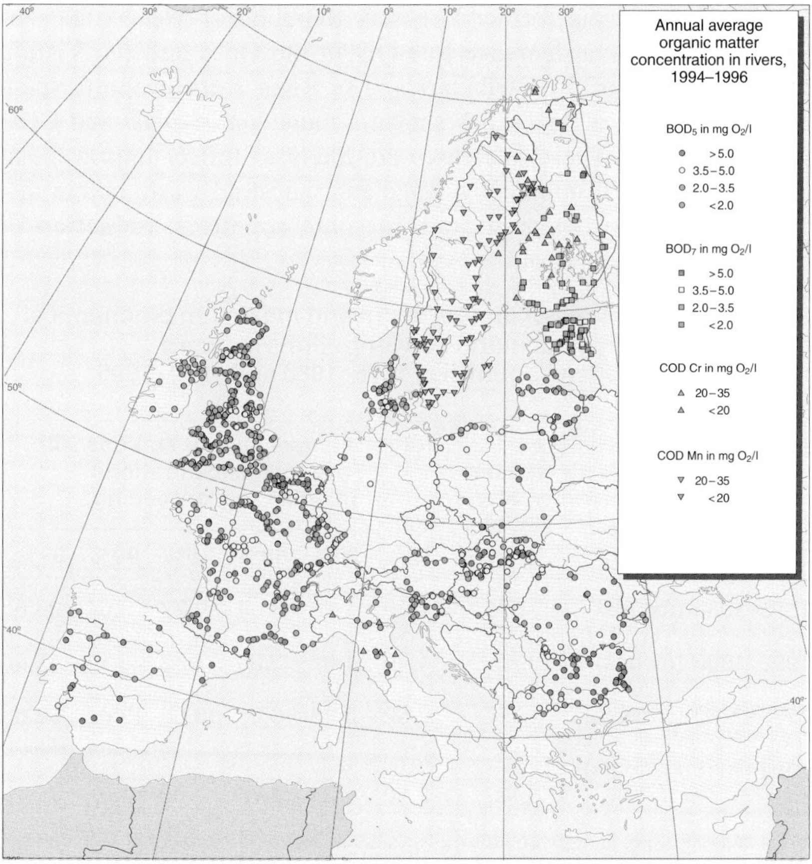

Source: European Environment Agency *(1)*.

Data on BOD and COD values in the former USSR show variation in different hydrographic regions. In the Baltic hydrological region, the Neman River is considerably polluted, with the trend for BOD and COD showing increased pollution in the late 1980s and 1990s compared with the 1970s. The River Lena in northern Siberia has high BOD and COD values, and no obvious decrease in organic pollution trends have been reported over time *(25)*.

The concentration of organic matter has decreased in some European rivers since 1981 (Table 2.3). The decrease has occurred particularly in the most polluted rivers, and the maximum BOD values recorded have declined. Significant reductions have been recorded in countries where the highest peaks were previously observed, such as Belgium, Bulgaria, the Czech Republic, Estonia, France, Hungary, Latvia and the former Yugoslav Republic of Macedonia. Poland also reports a decrease in BOD in the Oder and the Vistula rivers and most of their tributaries. The River Kura in the Caspian Sea hydrographical region has shown a reduction in BOD and COD since 1985 *(25)*. These decreases are likely to reflect improvement in the treatment of domestic sewage and industrial waste before discharge to the environment and, in some countries, reduction of

Table 2.3. Averages of annual mean and maximum biochemical oxygen demand concentrations in rivers in 29 European countries, 1975–1980 and 1992–1996[a]

	Number of stations	Percentages of river stations with oxygen concentrations (mg/litre) not exceeding:					
		10%	25%	50%	75%	90%	99%
1975–1980 (means)	575	1.40	1.96	3.04	4.77	7.54	26.8
1992–1996 (means)	1159	1.40	1.82	2.35	1.43	5.14	17.5
1975–1980 (maxima)	557	2.52	3.83	6.2	10.0	19.0	98.6
1992–1996 (maxima)	1407	2.50	3.24	4.75	7.40	11.2	39.1

[a] This table includes all representative river sites as reported by national authorities.

Source: European Environment Agency, unpublished data, 1998.

industrial or other economic activities caused by disruption of the economic system or by armed conflict.

Nitrate

About 80% of the nitrogen in rivers is present as nitrate *(6)*. In addition to being a nutrient, nitrate can have implications for human health if it is present in drinking-water at high concentrations (see page 36). In pristine rivers, the average level of nitrate has been reported to be about 0.1 mg per litre as nitrogen (mg/l N) *(26)* but, because of high atmospheric nitrogen deposition, the nitrogen levels of relatively unpolluted European rivers range from 0.1 to 0.5 mg/l *(6)*.

In a recent assessment of nitrate concentrations in European rivers *(1)*, 70% of the sites in the Nordic countries were reported to have concentrations below 0.3 mg/l N, whereas 68% of the sites in all European rivers were reported to have average annual nitrate concentrations exceeding 1 mg/l N in the period 1992–1996 (Table 2.4). About 15% of the sites had peak concentrations exceeding 7.5 mg/l N. The northern part of western Europe appears to had the highest concentrations, reflecting the intensive agriculture in these regions, although precipitation also greatly affects nitrate leaching (Fig. 2.4). High concentrations also occurred in eastern Europe. Most rivers in Belarus are reported to be contaminated *(27)*, whereas southern

Table 2.4. Averages of annual mean and maximum nitrate nitrogen concentrations in rivers in 30 European countries, 1975–1980 and 1992–1996[a]

	Number of stations	Percentages of river stations with nitrate nitrogen concentrations (mg/litre) not exceeding:					
		10%	25%	50%	75%	90%	99%
1975–1980 (means)	697	0.193	0.700	1.54	3.19	6.05	11.8
1992–1996 (means)	1525	0.193	0.720	1.73	3.53	5.89	9.78
1975–1980 (maxima)	685	0.392	1.23	3.12	5.66	11.4	24.4
1992–1996 (maxima)	1352	0.341	1.31	2.74	5.37	9.36	18.5

[a] This table includes all representative river sites as reported by national authorities.

Source: European Environment Agency, unpublished data, 1998.

Fig. 2.4. Average annual concentration of nitrate in selected European rivers, 1994–1996

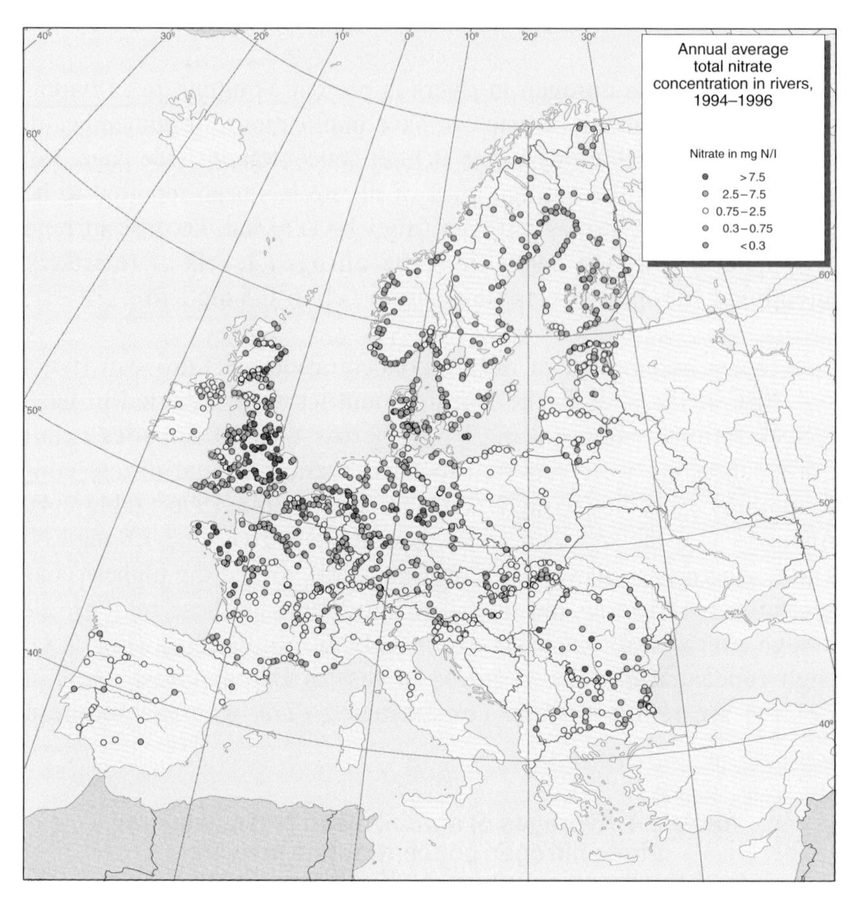

Source: European Environment Agency *(1)*.

European countries generally have lower concentrations. For many rivers in the former USSR, especially in basins of the Atlantic Ocean, and the Caspian and Aral Seas, concentrations of nitrates exceed natural concentrations *(28)*.

Phosphorus

Total phosphorus levels in undisturbed rivers are generally less than 25 µg/l, although natural minerals can contribute to higher concentrations. Phosphorus concentrations greater than 50 μg/l are attributed to human activities, and contamination resulting in levels higher than 100 μg/l may give rise to excessive growth of algae. A recent assessment of approximately 1000 European river sites demonstrated the extent of the influence of human activities on the phosphorus content of surface water, as only 10% of these rivers had mean total phosphorus concentrations not exceeding 50 μg/l (Table 2.5). Phosphorus concentrations were lowest in the Nordic countries, where 91% of sites have annual averages below 30 μg/l and 50% below 4 μg/l (Fig. 2.5) (although rising concentrations have been observed), reflecting nutrient-poor soil and bedrock, low population density and high rainfall. High phosphorus concentrations are especially found in a band stretching from southern England across central Europe to Romania and Ukraine. Western and eastern European countries exhibit very similar distribution patterns. Over the past two decades, concentrations of phosphorus have increased to 200 μg/l in the Dubasari water basin in the Republic of Moldova, causing eutrophication *(29)*.

Emissions of phosphorus from industrial regions of Denmark and the Netherlands have declined by up to 90% since the mid-1980s. As a result of the overall reductions in Europe, phosphorus concentrations in many rivers in the east and west of Europe generally decreased significantly between 1987–1991 and 1992–1996. The annual averages and maxima of total phosphorus and dissolved phosphorus

Table 2.5. Averages of annual mean dissolved (25 countries) and total phosphorus (24 countries) concentrations in European rivers, 1975–1980 and 1992–1996[a]

	Number of stations	Percentages of river stations with phosphorus concentrations (μg/litre) not exceeding:					
		10%	25%	50%	75%	90%	99%
1975–1980 (total)	105	86	150	317	683	1020	2834
1975–1980 (dissolved)	657	7	32	91	276	811	2832
1992–1996 (total)	546	50	100	172	290	576	2219
1992–1996 (dissolved)	1404	4	27	60	132	383	1603

[a] This table includes all representative river sites as reported by national authorities.

Source: European Environment Agency, unpublished data, 1998.

Fig. 2.5. Average annual concentration of phosphorus in selected European rivers, 1994–1996

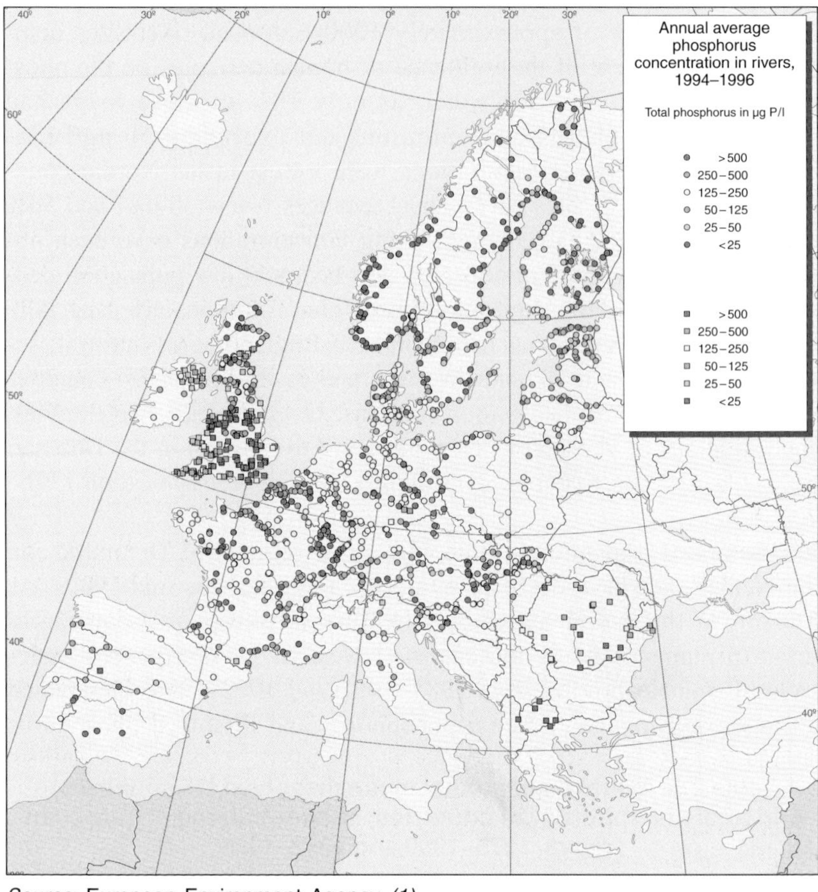

Source: European Environment Agency *(1)*.

exhibit the same patterns. The overall reduction in phosphorus emissions is likely to be especially caused by improved wastewater treatment and reduced use of phosphorus in detergents. The reduced pollution from point sources, however, means that contamination originating from diffuse sources, such as agriculture, is now relatively more significant. The trend in maximum values, however, suggests that excessive concentrations may be recorded even in generally improving sites.

In Belarus the amount of wastewater discharged from point sources declined by about 40% between 1991 and 1995, and the amount of inadequately treated water discharged to rivers declined by 75% *(27)*. In Ukraine, industrial wastewater discharge declined from 9813 million m^3 in 1992 to 7381 million m^3 in 1996 *(30)*.

Lakes and reservoirs

Europe has approximately 500 000 still bodies of water of over 1 hectare *(6)*. These comprise both natural lakes and artificial reservoirs. Limnicity (the total freshwater surface area of a region as a percentage of the total area of the region) is nearly always related to the density of lakes in the 10–100-km^2 range. Limnicity ranges from over 9% in countries such as Sweden *(31)* to about 1% in the United Kingdom and less than 0.5% in Greece *(32)*.

Lakes

Most countries have natural lakes, but their distribution is very uneven, with a large proportion concentrated in Finland, Norway, Sweden and parts of the Russian Federation. The former USSR had nearly 2.9 million lakes with a total surface area, including the Caspian Sea, of 892 850 km^2, about 4% of the total *(10)*. Iceland, Ireland and Scotland have significant numbers of natural lakes, but most of the largest European lakes are located in the Nordic countries and in the Alpine regions. In Albania, lakes occupy a surface area of 1150 km^2. Geological processes such as fluvial damming, volcanic activity and glacial events form natural lakes.

Reservoirs

Artificial reservoirs, usually formed by damming rivers, are constructed for a number of purposes, the most obvious being to compensate for spatial or temporal deficiencies in the natural water resource in relation to water demand. Reservoirs are built to provide water for irrigation, public supply and industrial use. Dams may also be built for the purposes of fisheries, electricity generation, flood control, low flow enhancement, transport, recreation or the storage of mining spoils *(33)*. The importance of the quality of the dammed water depends on the intended reservoir use, and is extremely important in reservoirs used for public supply, some industrial uses (such as food production), fisheries and recreation *(33)*. Many reservoirs have more than one purpose in practice, which can lead to conflicting priorities for different water uses. Fig. 2.6 compares population and total reservoir capacities in some European countries.

In absolute terms, Spain has the largest total major reservoir capacity (reservoirs deeper than 10 m) in the European Environment Agency's European Lakes, Dams and Reservoirs Database (ELDRED), representing over 50 000 million m³ of gross capacity or more than twice the total capacity of any other country except Norway (Fig. 2.7). In terms of gross capacity per inhabitant, Spain has just over 1000 m³ per head. This compares directly with an annual average renewal that is also estimated at just over 1000 m³ per head.

Spain and the United Kingdom have the largest number of reservoirs used for public water supply (approximately 300 and 400, respectively) and consequently may suffer problems of evaporation.

Fig. 2.6. Reservoir capacity per head of population for selected European countries

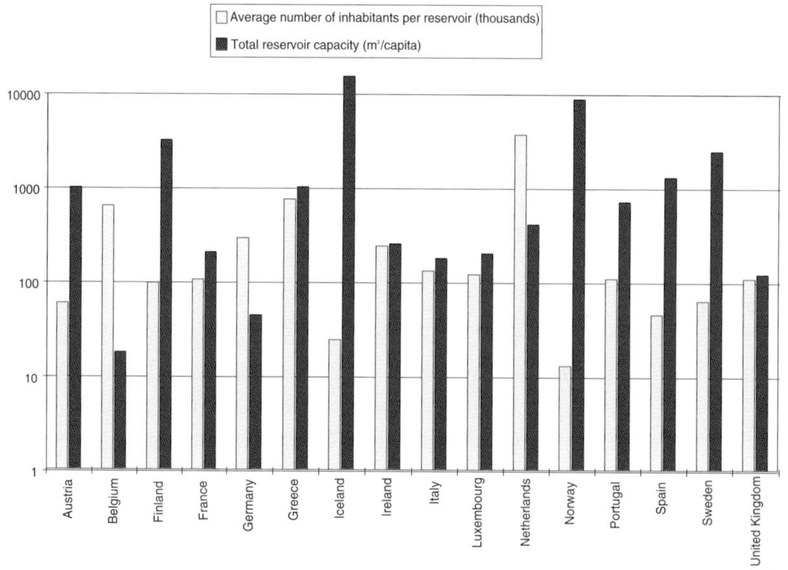

Average number of inhabitants per reservoir (thousands)
Total reservoir capacity (m³/capita)

Data sources: National Focal Points, ICOLD 1984/1988 and UN population statistics 1990, updated November 1997.
Note that storage data instead of total reservoir capacity data have been used for Norway.

Source: European Environment Agency *(34)*.

Fig. 2.7. Principal use of the major reservoirs in the European Lakes, Dams and Reservoirs Database (ELDRED)

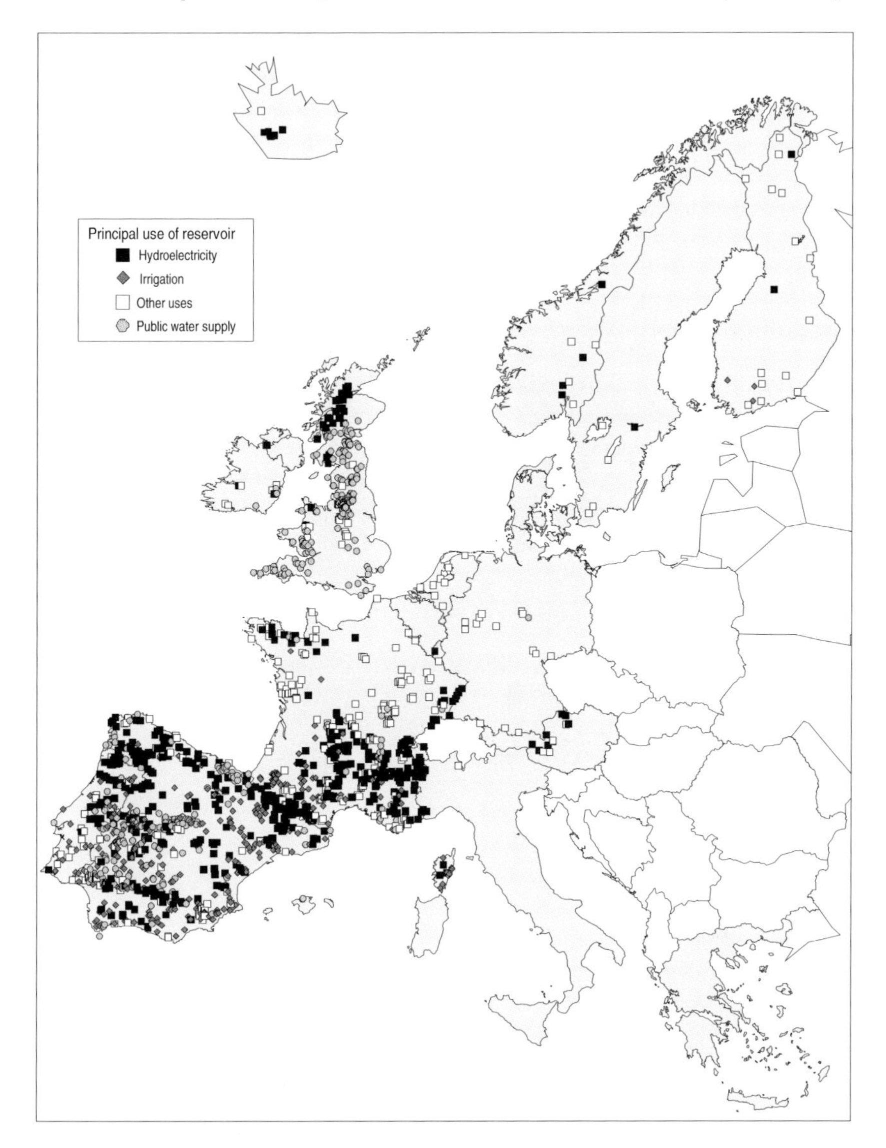

France, Germany and Italy also have many reservoirs. About 180 other major European reservoirs have public water supply as a secondary (or lower) priority (Fig. 2.8). Problems exist in identifying geological and geographical conditions suitable for sustainable

Fig. 2.8. Reservoirs and lakes in ELDRED used for public water supply

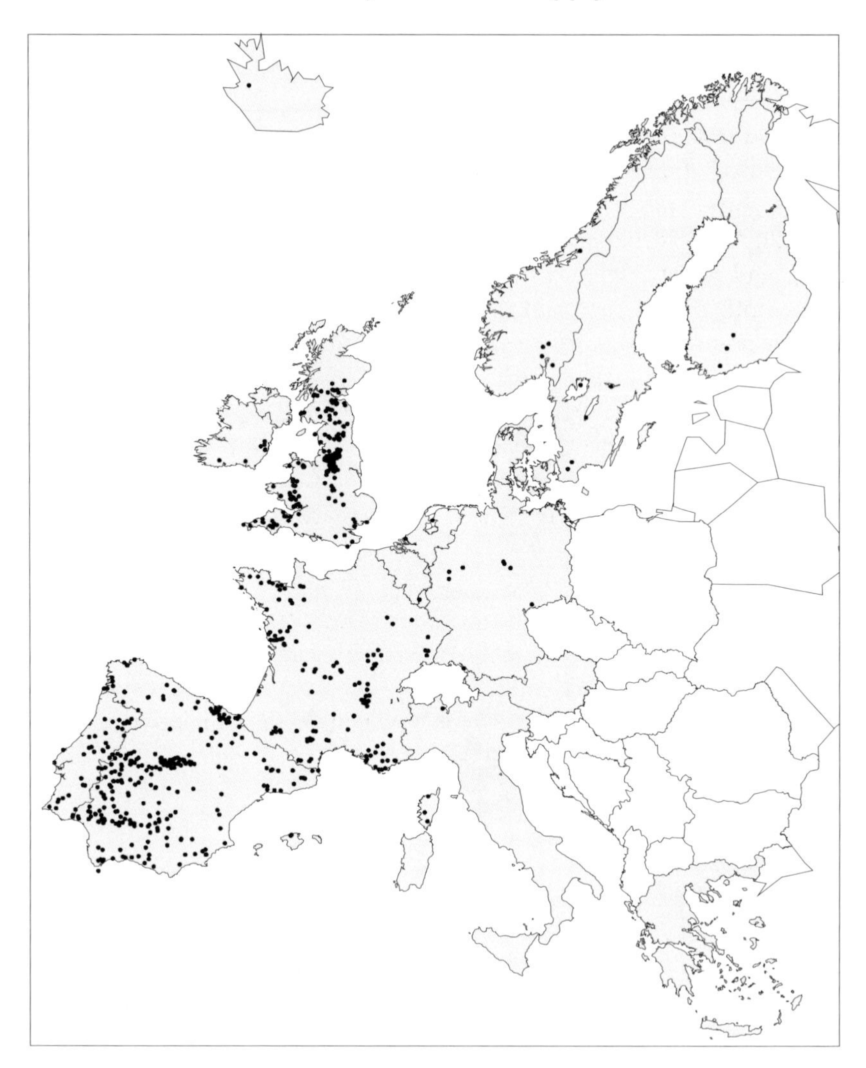

water management. In many countries agricultural activities create major quality problems, while in southern Europe problems exist because of high evaporation. The total capacity of European reservoirs used for public water supply (as their primary or lower-priority purpose) is about 32 000 million m^3, representing approximately 20% of total European reservoir capacity *(33,35,36)*. Many important public water supply reservoirs in Europe are relatively shallow (less than 10 m deep) and are therefore not part of the major reservoir data set contained in ELDRED; this may influence the apparent distribution of these reservoirs *(33)*.

Increasing reservoir capacity

The use of storage reservoirs overcomes the uneven distribution of natural water resources over time. Run-off in seasons of high rainfall can be held back to be used in drier seasons (seasonal regulation), and water available in wet years can be stored and used in dry years (interannual regulation). Increasing reservoir capacity is, therefore, a potential tool for meeting demand.

The reservoirs of Europe serve many functions other than providing drinking-water, such as electricity, irrigation, flood defence, recreation, navigation, fish farming and industrial supply (Fig. 2.7). Consequently, there is already a large storage capacity in a number of European countries.

Total reservoir capacity in Europe increased most between 1955 and 1985. The potential for the construction of further storage reservoirs in Europe is not likely to be large, since most suitable dam sites have already been selected and reservoir schemes implemented. Consequently, future dams will face higher economic and environmental costs *(33)*.

Water quality

Water quality is a prime consideration in lakes and reservoirs used for public water supply. Such bodies of water are often more vulnerable and sensitive to pollution than watercourses or marine waters, since water volumes are not frequently renewed and lake morphology tends to lead to accumulation of pollution. Acidification, resulting from atmospheric deposition, and eutrophication caused by excessive nutrient loading are the main problems affecting the quality of European lakes (Box 2.2). Only lakes in sparsely populated areas or mountainous regions away from populated areas have low nutrient levels. In densely populated areas such as western and central Europe, many lakes have elevated phosphorus levels as a result of human activities (Fig. 2.9). Lakes and reservoirs situated in lowland regions are most likely to be subject to higher nutrient loads.

Box 2.2. Eutrophication

Excessive discharges of nutrients (phosphorus and nitrogen) to lakes and reservoirs cause an imbalance in the aquatic ecosystem. An N : P ratio of about 10 : 1 by weight is considered to be ideal for algal growth. In fresh water (N : P ratio > 10 : 1), phosphorus is naturally the limiting nutrient. The maximum permissible phosphate loads per surface area for different types of water body have been published. Phosphorus concentrations above 10–20 µg/l result in algal blooms that cause a variety of problems. In marine water, nitrogen is most often considered to be limiting (N : P ratio < 10 : 1) *(38)*. Both nutrients can be limiting on a seasonal basis – nitrogen in summer and phosphorus in winter – although these nutrients are only truly limiting to algal growth at very low concentrations. The threshold concentration of nutrients above which eutrophication becomes a problem depends on the topography and the physical and chemical nature of the water. Restricting phosphorus discharge to surface water is the only way to successfully control eutrophication.

Increased standing crops of algae reduce light penetration through the water column, thereby reducing the depth at which rooted higher plants can grow. Thus, lakes tend to be dominated either by rooted macrophytes (shallow or nutrient-poor lakes) or by algal growth in the water column (deep or nutrient-rich lakes). Raw water containing high levels of algae requires additional treatment if it is to be used for potable supply, because of filter blockage, filter penetration, polysaccharide production (which increases dissolved organic carbon levels and interferes with the stability of the floc blanket), generation of taste and odour, and toxin production.

The high biological productivity in nutrient-enriched water means that BOD and sediment oxygen demand in the water column are high as dead material is broken down. Thus, oxygen can be stripped out of the water column if the body of water is not well mixed. This can result in a series of ancillary problems with raw water quality, such as high manganese, iron, ammonia and hydrogen sulfide concentrations, all of which are released from the sediment under reducing conditions.

Artificial mixing or aeration of lakes and reservoirs is widely used to prevent these ancillary problems. In some cases, reservoirs may be allowed to thermally stratify and bottom water may be selectively removed from the reservoir – either by opening the scour valve at the base of the dam or by pumping the bottom water over the dam. In deep lakes and reservoirs, artificial mixing may reduce algal levels by circulating algae from well lighted surface waters to a depth at which there is insufficient light for net photosynthesis.

Such reservoirs are often used for irrigation or public supply and are particularly sensitive to eutrophication *(33)*.

The release of nutrients (particularly nitrogen and phosphorus), which predominantly originate from agriculture and domestic sewage, can significantly affect the balance of the aquatic ecology and often produce algal blooms throughout the body of water and at the surface.

Fig. 2.9. Distribution of phosphorus concentrations in selected European lakes and reservoirs by country

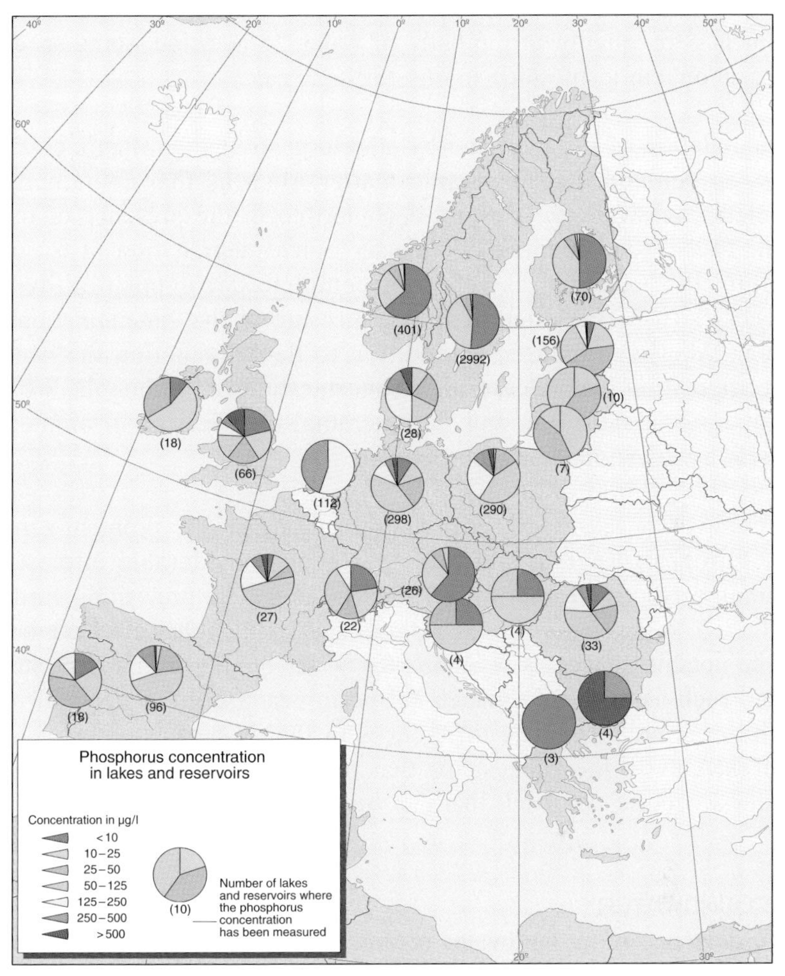

Number of lakes per country: Austria 26; Belgium 4; Denmark 28; Estonia 156; Finland 70; France 27; Germany ~300; Hungary 4; Ireland 18; Latvia 10; Lithuania 7; Netherlands 112; Norway 401; Poland 290; Portugal 18; Romania 33; Slovenia 4; Spain 96; Sweden 2992; Switzerland 22; the former Yugoslav Republic of Macedonia 3; United Kingdom 66.

Source: European Environment Agency *(1)*.

The resultant cloudy water requires additional treatment to make it fit for consumption, and algal by-products interfere with water treatment. As the algae die and fall to the bottom of the lake or river, their decay exerts an oxygen demand that can deplete the dissolved oxygen in the water column (Box 2.2). Significant amounts of phosphorus can be contained in sediment, and its release can significantly influence the quality of the body of water. In addition, some species of cyanobacteria produce toxins *(37)*.

The pollution of surface water by other organic matter often increases oxygen demand. Much organic matter, such as that contained in domestic wastewater, is easily decomposed in the presence of oxygen. The oxygen required for the biological degradation of organic pollutants (measured as BOD) or complete decomposition by all processes (COD) is often used as a measure of the contamination of water by organic matter. In addition to de-oxygenating the water, decomposition can release high concentrations of ammonia, which is toxic to aquatic life, and the microbial nitrification of ammonia exerts an additional oxygen demand.

Eutrophication has become a widespread and severe problem across the European Region, and there has been considerable interest in finding methods to improve affected lakes and in preventing further pollution by nutrients (Boxes 2.3 and 2.4). Although there have been some notable successes in improving water quality by nutrient stripping, such as at the Wahnbach Talsperre near Berlin, successful restoration is not easy. The differences in the properties of lakes and the changes in the ecosystems that result from eutrophication preclude a generic approach that can be applied to all lakes.

Groundwater

The presence of groundwater resources depends largely on the geology of a country. The arrangement of permeable and impermeable layers of rock and impermeable glacial or glaciofluvial deposits and the presence of underground caverns determine the water storage capacity. Groundwater is recharged primarily by percolation of water through the soil.

Groundwater is generally of higher quality than surface water since it is, except in some karstic horizons, less vulnerable to anthropogenic contamination. Less treatment is therefore required to make it safe for use as drinking-water. Nevertheless, groundwater is

Box 2.3. Restoration of eutrophic lakes

Strategies to reduce the nutrient loading of lakes include measures to reduce levels of nutrients entering the tributary streams. These might target diffuse sources by, for example, reducing fertilizer application rates or the density of animal stocks. However, greater success has been achieved to date by tackling point sources, such as by installing nutrient stripping at wastewater treatment plants or by reducing the phosphorus content of detergents. Alternative (but much less common) approaches include stripping nutrients from reservoir feeder streams before the water enters the lake, or circulating nutrient-rich reservoir water through a phosphorus stripping plant and returning the treated water to the reservoir. Metal salts, such as ferric sulfate, can be dosed to the lakes as flocculants to precipitate the dissolved phosphorus, but the heavy metals build up in the sediment where they may have a toxic effect.

Unfortunately, water quality is not likely to improve and the ecosystem is not likely to be restored immediately after any of these measures. Phosphorus stored in the sediment of lakes is released into the water column, and high dissolved concentrations can result from this internal source. Even when this has stabilized at a lower rate, the ecological balance of the lake may have been altered so that, even if phosphorus concentrations are reduced to one third of their previous level, algal densities may show little change. To reduce this lag, many lake restoration schemes involve either sediment removal or sediment sealing, or inactivation to prevent or inhibit phosphorus release.

Because eutrophication changes ecosystems, the aquatic life in the lake may need to be manipulated in addition to addressing the nutrient loading directly. This appears to be especially true for shallow lakes. Removal of the fish species that prey on zooplankton such as *Daphnia* spp. is a commonly used strategy, thereby increasing the standing crop of zooplankton and increasing grazing pressure on the algal population.

There are numerous examples of restoration schemes for lakes or reservoirs, such as Wahnbach Talsperre (Box 2.4), Finjasjön and Lake Geneva *(39,40)*. The success of different approaches and individual restoration programmes varies greatly. The economic and ecological effects of eutrophication are high, in terms either of additional water treatment or of restoration schemes. Prevention is better than cure.

susceptible to contamination by certain chemicals, with particular problems occurring from nitrate, pesticides and volatile organic solvents. The virtual lack of any losses from volatilization, often minimal biodegradation and a long recharge time mean that groundwater sources are very slow to recover from contamination.

Abstraction from groundwater

Aquifers can be an efficient natural solution to seasonal water scarcity. Groundwater acts as a year-round resource and, if recharge during wet periods is sufficient, can be used to supply water during times of low precipitation. Water quality is often good, and aquifers

Box 2.4. Controlling eutrophication – case studies

Lake Zurich Lower Basin, Switzerland
Problem. Improvements to wastewater treatment facilities were originally concentrated on reducing organic loadings. Nutrient enrichment was then considered a target for action, although Secchi depths were quite high at 3–11 m.

Action undertaken. Phosphorus removal has been progressively undertaken at wastewater treatment plants in the catchment area since the 1970s. A national ban on phosphates in detergents was introduced in 1986.

Results. Little change in the Secchi depth of phytoplankton levels despite substantial reductions in phosphorus levels.

Lessons. Reducing the in-lake phosphorus concentrations by a factor of about 3 from 1974 levels has had relatively little effect on standing populations of algae.

Wahnbach Talsperre, Germany
Problem. The reservoir received strongly increased phosphate loadings in the 1960s and became eutrophic, with blooms of cyanobacteria.

Action undertaken. From 1969 the deep part of the lake received artificial aeration. A full restoration programme was started in 1977. This involved removing phosphorus from the tributary by precipitation with iron (III) chloride and subsequent filtration of the precipitate.

Results. Nutrient removal of the feeder river removed most incoming particulate phosphorus, which is mainly of mineral composition and would otherwise have significantly contributed to the fixation of phosphorus in the sediment. After about 5 years, measurements of chlorophyll concentration and opacity of the lake water indicated that the restoration has been successful in decreasing the algal population. Measurements of phosphorus concentrations confirm a successful reduction of nutrient concentrations.

Lessons. The principal populations of cyanobacteria decreased very rapidly after the onset of the phosphate removal, but another mobile species was able to utilize the phosphorus in the upper layers of the sediment for about 5 years after the restoration started, until this phosphorus was also depleted.

Sources: Klapper *(41)*; Ryding & Rast *(39)*; Sas *(40)*.

can provide quality reserves in areas where surface runoff in summer proves insufficient to maintain acceptable standards of water quality. The proportion of freshwater demand from groundwater abstraction varies from country to country. In the EEA area, about 18% of the total water abstraction is from groundwater, ranging from 91% in Iceland to less than 10% in Belgium. Groundwater abstraction accounts for 90% of total freshwater demand in Georgia *(42)* and 57% of total abstraction in Belarus *(27)*. The use of aquifers depends on annual recharge and requires effective management if sustainability is to be guaranteed. In countries with sufficient groundwater reserves (Austria, Denmark, Iceland, Portugal and Switzerland) more than 75% of the water for public supply is abstracted from groundwater. In other countries with scarce groundwater reserves or with abundant, clean, surface water reserves the proportion abstracted for drinking-water falls to below 50%.

Over-exploitation and saline intrusion

Aquifers vary in size, and many of those exploited for abstraction contain large volumes of water. However, aquifers usually recharge slowly, and current abstraction levels may, in some cases, not be sustainable. In many cases, determining whether over-exploitation is occurring is difficult. It is often thought of as being a relatively straightforward question of water taken out of the aquifer and water filtering back into it. Difficulties in estimating long-term recharge confound this simple approach.

Exploitation of groundwater sources beyond a sustainable level can affect the environment (such as loss of wetlands and effects on river ecosystems) and reduce the future availability of the resource. When aquifers near the coast are over-exploited, salt water may intrude, reducing the quality of the groundwater. In some coastal regions in southern Europe in particular, aquifers have very limited annual recharge and saline intrusion has occurred *(1)*, reducing the flexibility of water resources for required uses. Over-exploitation is most likely to occur in arid or semi-arid regions where recharge is low, but there are nevertheless a number of over-exploited aquifers in temperate climates. The rural aquifer north of Nottingham in the United Kingdom is experiencing over-exploitation, causing wetlands to dry out. Irregular surface water resources and an increasing water demand have led to dependence on groundwater. Aquifers supplying Barcelona, Marseilles, Athens and the French Riviera, for example, are already stressed and are expected to deteriorate further *(43)*.

Aquifer recharging

Artificial recharging of aquifers is used to avoid or rectify problems of over-exploitation. Surface water and wastewater are both potential sources of water for artificial recharging. Direct injection of water into the groundwater zone is one method of recharging and can be used to create a barrier against saltwater intrusion in coastal areas. Surface spreading or dune or bank filtration are more often used; as the water percolates downwards, the soil or sand acts as a natural filter to remove particulate matter, microorganisms and some dissolved compounds. Depending on the soil characteristics and the extent and type of contamination of the water used for recharging, the groundwater may become polluted. Where reclaimed wastewater is used, treatment may be necessary to eliminate microorganisms.

In the Netherlands, where depletion of groundwater resources is recognized as a problem *(44)*, a significant proportion of the water abstracted from surface sources is used to artificially recharge groundwater using such methods as dune filtration. Artificial recharging is also common in Germany, where water is treated (such as by coagulation) to reduce contamination by microorganisms and organic compounds before it is used for recharging.

Bank-filtered water may also be used for drinking-water. In this case, water from a contaminated surface water source (usually a river) is allowed to filter into the groundwater zone through the river bank and to travel through the aquifer to an extraction well some distance from the river. In some cases there is a very short residence time in the aquifer, perhaps as little as 20–30 days, and there is almost no dilution by natural groundwater *(45)*.

Nitrates

The natural level of nitrate in groundwater is generally below 10.0 mg/l. Elevated nitrate levels are especially caused by the agricultural use of nitrogen fertilizers and manure in excess of plant requirements, or by their application at the wrong time of the year. Local pollution from municipal or industrial sources can also be important. On-site sanitation and leaky sewer pipes in urban areas may also contribute to increased nitrate levels. Certain types of aquifer, such as alluvial and shallow aquifers, are more vulnerable to nitrate pollution than others because of differences in such features as hydrogeology and land use. Deep or confined aquifers are generally better protected.

Nitrate moves relatively slowly through the ground, so there can be a significant time lag between the polluting activity and the detection of the pollutant in groundwater (typically between 1 and 20 years). The general intensification of agriculture over the last

30 years has only relatively recently been reflected in increasing nitrate concentrations in groundwater.

The WHO guideline value for nitrate in water intended for human consumption is 50 mg/l NO_3 *(46)*. The EU Drinking Water Directive *(47,48)* permits the same concentrations. The results from groundwater monitoring programmes in 17 European countries showed high levels of nitrate (greater than 25 mg/l) in groundwater in 50% of the sampling sites in Slovenia. In eight countries this level was exceeded in about 25% of the sites and, in one country (Romania) 35% exceeded 50 mg/l *(1)*. In Denmark, for example, about 2% of all supply wells have been closed since 1986 because nitrate concentrations were above 50 mg/l. Fig. 2.10 provides an overview of the regions in Europe where groundwater is affected by high nitrate concentrations.

High nitrate concentrations can be localized, largely depending on land use. This is the case in Latvia, for example, where groundwater is polluted by agricultural land regionally. At sites with intensified use of agricultural chemicals, nitrate concentrations increased in 55% of boreholes between 1988 and 1990. In the newly independent states of the former USSR, 20% of groundwater sites are contaminated with agricultural contaminants *(7)*. The Republic of Moldova reports that half the drinking-water supplies from groundwater in the Prut River basin have nitrate concentrations exceeding 45 mg/l *(29)*. High nitrate concentrations (in excess of 50 mg/l) are found in 15% of sampling sites in Austria, and in 28% of sites more than 30 mg/l was found.

In many countries, nitrate concentrations in groundwater have increased since the middle of the 20th century, following the intensification of farming methods and an increase in the area used for farming. Nitrate levels are currently above maximum permissible levels in 76% of wells in Belarus, with concentrations up to 300–600 mg/l *(27)*. In a number of countries in the eastern part of the Region, such as Hungary and Romania, nitrate levels have declined. This is probably related to a reduced economic potential for purchasing agricultural chemicals. Monitoring data show different trends in a number of western European countries, even over a short period in the 1990s, although a much longer period of time is needed to

Fig. 2.10. Regions in selected countries in Europe affected by high nitrate concentrations in groundwater

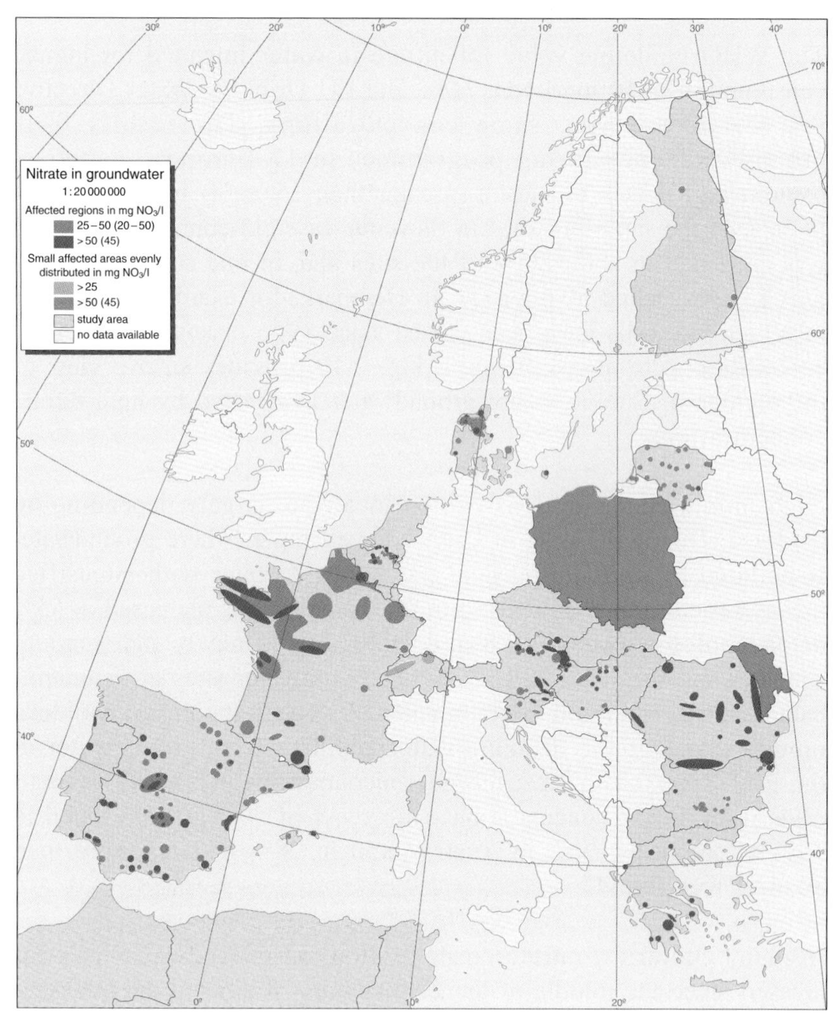

Source: European Environment Agency *(1)*.

establish meaningful trends and to eliminate the influence of short-term changes such as rainfall patterns (Table 2.6). Nitrate concentrations in some countries do not appear to have increased over this short time, possibly because of increased awareness and policies to reduce the use of nitrate fertilizers.

Table 2.6. Percentage of groundwater monitoring stations in selected European countries with nitrate concentrations that increased, were unchanged or decreased from the early 1990s to the mid-1990s

Country	Number of sites	Increased (%)	Unchanged (%)	Decreased (%)
Austria	979	13	72	15
Denmark	307	26	61	13
Finland	40	27	43	30
Germany	3741	15	70	15
United Kingdom	1025	8	80	12

Source: European Environment Agency, unpublished data, 1998.

Pesticides

Pesticides are a large and diverse group of chemicals, with different physicochemical properties and toxicity, that are used in a wide range of applications. These include agriculture, horticulture and public amenities, public health in the control of vectorborne disease and, in small amounts, in the home. They can enter surface water and groundwater from point sources (disposal or spillage) and diffuse sources (used in agriculture and amenity use). The types and amounts of pesticides found in raw water sources used for drinking-water depend on many factors. Examples include the physicochemical properties of the chemical (affecting the extent to which it binds to soil or leaches), its biodegradability, the soil type, the geological characteristics of the underlying rock, the weather (especially precipitation and the soil moisture content) and the time of application (application before rain makes run-off more likely). The most important factor is the quantity of pesticides used in the catchment area. Thus commonly used herbicides such as the triazines (atrazine and simazine) and the urons (diuron and chlortoluron), which are used in relatively large quantities, are often reported as occurring in raw water sources. Breakdown products such as desethylatrazine are also detected.

Pesticides applied in agriculture can filter down to the groundwater under normal field conditions although the type of crop, the method and rate of applying the pesticide and the equipment used may influence the infiltration rate. During application in the spring and subsequently through the summer months, the moisture deficit within the soil profile restricts the vertical movement of pesticides through it. Adsorption onto organic carbon further retards movement through the profile, and degradation by soil microbial processes reduces the concentration of pesticides available for leaching. In wetter months water will start to migrate through the unsaturated zone. The solute will be subject to further biochemical decay as it moves through the aquifer matrix, moving slowly via intergranular flow paths. Where there are karstic conditions (fissure flow) or high flux rates due to blind ditches or topography, however, groundwater recharge pathways tend to be highly developed and, under these conditions, pesticides have the potential to migrate through thick layers of unsaturated material to infiltrate groundwater quickly. Infiltration of pesticides can occur year round following heavy rainfall, especially when this follows soon after application.

Because of the huge number of pesticides (over 800 are approved for use in Europe), efficient monitoring of residues in the environment is complex and expensive. The most cost-effective basis for a monitoring programme is likely to be restricting analysis to the pesticides most likely to be used in the area. However, this approach renders difficult the comparison of data from different sites and countries and at different points in time. Such differences in reporting concentrations as exceeding a specific limit or being above the limit of detection also make comparison difficult. In addition, monitoring may concentrate on sites that are suspected or known to be contaminated, therefore producing a non-representative sample of monitoring results. Table 2.7 provides an overview of selected pesticides at the country and regional levels and the percentage of sampling sites where the average annual pesticide concentration exceeds 0.1 μg/l.

Much of the monitoring of pesticides in groundwater in EU countries has been carried out solely to comply with the EU Drinking Water Directive *(47,48)*. Current monitoring is not sufficient to establish the extent of contamination of groundwater with pesticides or to establish any trend in concentrations *(49)*.

Table 2.7. Percentage of sampling sites (total number in parentheses) in selected European countries with average annual pesticide concentrations greater than 0.1 μg/l[a]

	Austria	Czech Republic	Denmark	France	Germany	Luxembourg	Norway	Republic of Moldova	Romania	Slovakia	Slovenia	Spain	United Kingdom
Atrazine	16.3 (1 666)		0 (625)	*b*	4.1 (11 690)	*b*					32.1 (84)		*b*
Simazine	0.2 (1 248)		0.3 (625)	*b*	0.9 (11 630)	*b*					4.8 (84)		
Lindane		0 (215)		*b*	*b*					25 (8)		*b*	
Desethylatrazine	24.5 (1 666)				7.1 (11 690)						47.6 (84)		
Heptachlor				*b*						0 (12)		*b*	
Metolachlor	1.1 (1 248)					*b*					4.8 (84)		
Bentazone						*b*	80 (5)						
DDT		0 (215)								0 (12)			
Dichlorprop			0.6 (623)				83.3 (6)						
Methoxychlor		0 (206)								8.3 (12)			
Desisopropylatrazine	1.3 (1 666)												
Bromacil					3.5 (6 650)								
DDE, DDD, DDT								*b*					
DDD (*p, p'*), DDT (*p, p'*)												*b*	
Chlortoluron													*b*
Dichlorbenzamide			13.7 (102)										
Dieldrin				*b*									

Table 2.7. (contd)

	Austria	Czech Republic	Denmark	France	Germany	Luxem-bourg	Norway	Republic of Moldova	Romania	Slovakia	Slovenia	Spain	United Kingdom
Diuron													*b*
Endosulfan I												*b*	
Endosulfan sulfate												*b*	
GCCG-a,b								*b*					
Hexachlorocyclohexane, α, β, γ												*b*	
Hexachlorobenzene										0 (10)			
Hexazinon					*b*								
Isoproturon													*b*
Linuron													*b*
4-Chloro-2-methylpheno-oxyacetic acid							100	(2)					
Mecoprop			0.2 (625)										
Metalaxyl											*b*		
Metazachlor						*b*							
Parathion-methyl												*b*	
Pentachlorophenol		0 (207)											
Phosphamide								*b*					
Phozalone								*b*					
Prometryn											2.4 (84)		
Propazine					0.6 (10 890)								
Sum (hexachlorocyclohexane)									*b*				
Sum (hexachloro-cyclohexane + DDT)									*b*				

[a] The numbers indicate the percentage of sampling sites with concentrations > 0.1 μg/l and the number of sampling sites in parentheses.

[b] Data available at the regional level only.

Source: Scheidleder et al. *(33)*.

The pesticides that have been detected most frequently in groundwater are the triazine herbicides, especially atrazine and simazine, and their breakdown products. These are broad-spectrum herbicides that have been used extensively in both agricultural and non-agricultural situations. Because they frequently appear in groundwater, several countries have banned or restricted the use of products containing these active ingredients, and a recent assessment *(49)* revealed a statistically significant downward trend in the contamination of groundwater with atrazine and its metabolites in a number of countries, including Austria and Switzerland and parts of France, Germany and Latvia. In Baden-Württemberg in Germany, however, where atrazine concentrations in groundwater appear to be decreasing, concentrations of another triazine herbicide, hexazinon, show an upward trend *(49)*.

Limited data are available on contamination of surface water with pesticides. On the island of Fyn in Denmark, for example, water samples taken in 1994 and 1995 from six streams showed 25 different substances in concentrations exceeding the detection limit. The highest concentrations were found in the spring, coinciding with pesticide application in the fields *(1)*.

Hydrocarbons and chlorinated hydrocarbons

Hydrocarbons and chlorinated hydrocarbons are important contaminants of groundwater in a number of European countries. Chlorinated hydrocarbons are widely distributed in groundwater in western Europe, where they have been extensively used as solvents. Because of their volatility, chlorinated solvents that are accidentally released to surface water may be rapidly removed by evaporation and are not a significant pollution problem. In contrast, contamination of groundwater is extremely persistent. A number of studies have investigated the extent of contamination of aquifers by chlorinated organic solvents, such as trichloroethene and tetrachloroethene, in western Europe. The picture that emerges is of widespread, low-level contamination in aquifers under industrialized areas, with localized high levels of contamination.

Groundwater contamination by hydrocarbons from petrol stations, petrochemical plants and pipelines and military sites, in particular, is a problem throughout Europe. Groundwater in Estonia, for example, is polluted by leakage from fuel tanks *(50)*.

TRANSBOUNDARY WATERS

Many countries share water resources with a neighbouring country. Still more receive imported water in the downstream flow of rivers. Most of the major European river basins are transboundary in nature. For example, the Danube passes more than ten countries between its source in Germany and the Black Sea.

Some countries are highly dependent on transboundary flows and, for some countries, water originating outside the country is essential to meet the needs of the population. The Netherlands, for example, abstracts only 16% of the total available resources, but this is equivalent to over 100% of the water that originates within the country (Table 2.8). Over 95% of the total fresh water in Hungary originates outside the country, as do over 80% of the resources of Slovakia. These countries are therefore especially vulnerable to the effects of abstraction, impoundment and pollution by countries upstream.

Several international agreements cover the management of shared water resources in Europe. These include the 1992 ECE Convention on the Protection and Use of Transboundary Watercourses and International Lakes (Box 2.5) *(51)*, as well as a number of action plans and conventions relating to specific rivers and bodies of water. Nevertheless, the potential for conflicts of interest exists, and there are problems relating to the contamination of international river

Table 2.8. Water use intensity calculated using abstraction as a percentage of total available resources and internally generated water resources in selected European countries

Country	Water use intensity (%) Total available resources	Internally generated water resources
Belgium	72	91
Bulgaria	6	46 – > 100
Lithuania	19	31
Hungary	5	96
Netherlands	16	136
Portugal	10	26
Republic of Moldova	13	89
Romania	9	49

Source: European Environment Agency *(6)*.

Box 2.5. The 1992 Convention on the Protection and Use of Transboundary Watercourses and International Lakes

Objectives

- To prevent, control and reduce pollution of waters causing or likely to cause transboundary impact.
- To ensure that transboundary waters are used with the aim of ecologically sound and rational water management, conservation of water resources and environmental protection.
- To ensure that transboundary waters are used in a reasonable and equitable way, taking into particular account their transboundary character, in the case of activities that cause or are likely to cause transboundary impact.
- To ensure conservation and, where necessary, restoration of ecosystems.

Actions achieved

- Measures required for preventing, controlling and reducing water pollution.
- Ratified by 31 Parties (30 countries plus the European Union) as of 28 September 2000.
- Convention came into force on 6 October 1996.

Source: United Nations Economic Commission for Europe *(51)*.

catchment areas and aquifers. Important international monitoring programmes also exist for large lakes where international commissions have been set up to coordinate action programmes. Notable examples of such programmes are:

- Lake Geneva – France and Switzerland (protection, navigation, monitoring and abstraction);
- Lake Constance – Austria, Switzerland and Germany (protection and abstraction);
- Inari – Finland and Norway (regulation of hydroelectric power);
- Lugano – Italy and Switzerland (regulated by the Convention) *(1)*; and
- GEMS/Water – an international programme on water quality monitoring and assessment.

The last mentioned is jointly implemented by WHO, the World Meteorological Organization, the United Nations Environment Programme and the United Nations Educational, Scientific and Cultural

Organization. It aims to assist countries in establishing and strengthening their water quality monitoring operations, and provides methodological and quality assurance support to the sustainable management of freshwater resources.

Effects of climate change on the quality of water resources

Saline may intrude into groundwater supplies as a result of a rising sea level. Water quality will be most affected where salinity is already a problem because aquifers are being over-exploited *(22)*. Less obvious effects of a rising water table may include the release of contaminants from septic systems and of pollutants from underground waste-disposal sites. Coastal flooding is likely to be the most immediate and significant short-term effect, however, causing disruption of the sanitation infrastructure and potential contamination of watercourses by sewage.

The predicted increase in episodes of heavy rainfall will result in increased contamination of surface water by runoff, especially by eroded soil, microorganisms, pesticides and fertilizers *(22)*. Areas with no vegetation cover are especially vulnerable to run-off of soil and particulate matter during heavy rainfall.

The resulting poorer water quality will require more robust water treatment measures, although the effectiveness of water treatment may also be compromised by some of the predicted changes. Heavy rainfall and high winter river flows will result in source water with less dissolved salts and lower alkalinity. Greater variability of river flows (increased winter flow, reduced summer flow and increased variability between years) will make it more difficult to plan and design appropriate facilities for treating and distributing drinking-water.

3

Driving forces and pressures on water resources

The water resources of a country depend on many factors, including its climate, geography and geology. The pressures on these resources are determined by the population density and the agricultural, industrial and domestic practices of the population.

The water resources of the European Region as a whole are sufficient to supply the requirements of the population. Comparisons of total freshwater abstraction with the resources available suggest that most European countries have sufficient resources to meet the national needs (Fig. 3.1 and 3.2). Nevertheless, water resources are unevenly distributed and are reported to be insufficient in southern Ukraine, the Republic of Moldova, the middle and lower reaches of the River Volga, the Caspian lowland, the southern parts of western Siberia, Kazakhstan and the Turkmenistan lowland *(7)*.

Data relating to the relative demands exerted by different sectors on a country's water resources vary significantly, depending on the source of the data *(52)*. These differences largely result from differing approaches to data collection (such as including industries supplied by the public water networks in statistics on municipal water use, and including cooling water in power plants or water used for hydroelectric power production in the definition of industrial use). Nevertheless, in many cases, especially in southern Europe (Greece, Italy, Portugal and Spain), agricultural water use (mainly for irrigation and livestock) is highly significant and can account for more

Fig. 3.1. Abstraction of fresh water in selected European countries as a percentage of total renewable water

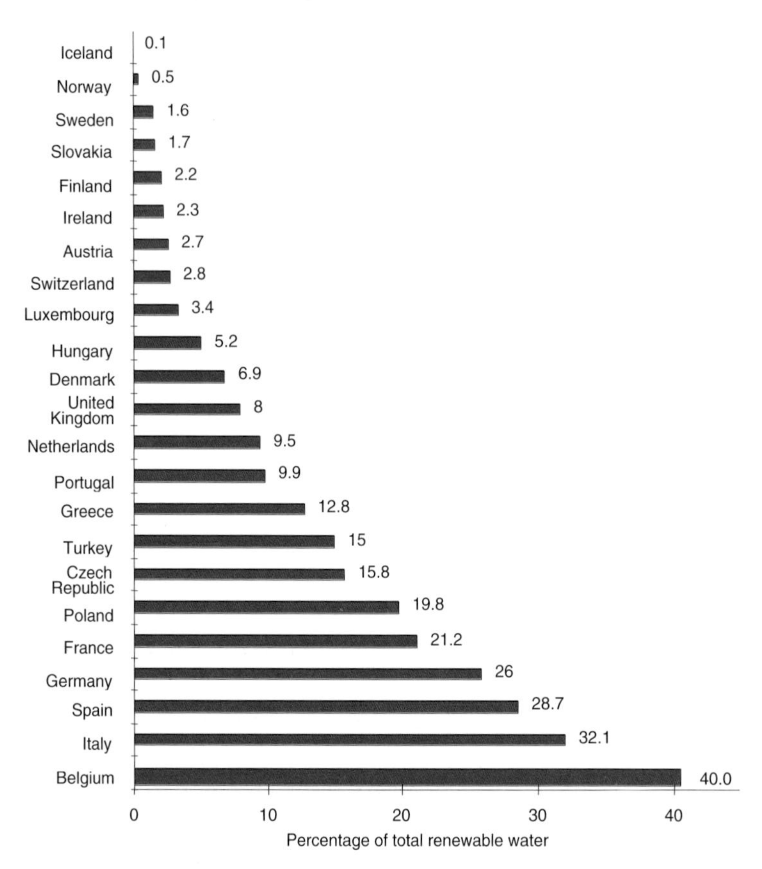

Source: Krinner et al. *(52)*.

than 60% of total abstraction (Fig. 3.2). In the eastern part of the Region, agricultural water demand has declined since 1988 because of changes in land ownership *(1)*. The Russian Federation showed an overall reduction of 0.7% in abstraction for irrigation between 1993 and 1996, and the water used for irrigation in the Republic of Moldova declined between 1990 and 1994. However, water use for irrigation has increased in Kazakhstan, Turkmenistan and Uzbekistan. In these countries about 50% of total water demand is for irrigation *(7)*. Uncontrolled development of irrigation can severely affect water resources, the Aral Sea being the most dramatic example (Box 3.1).

Fig. 3.2. Sectoral use of water in the countries of the European Union

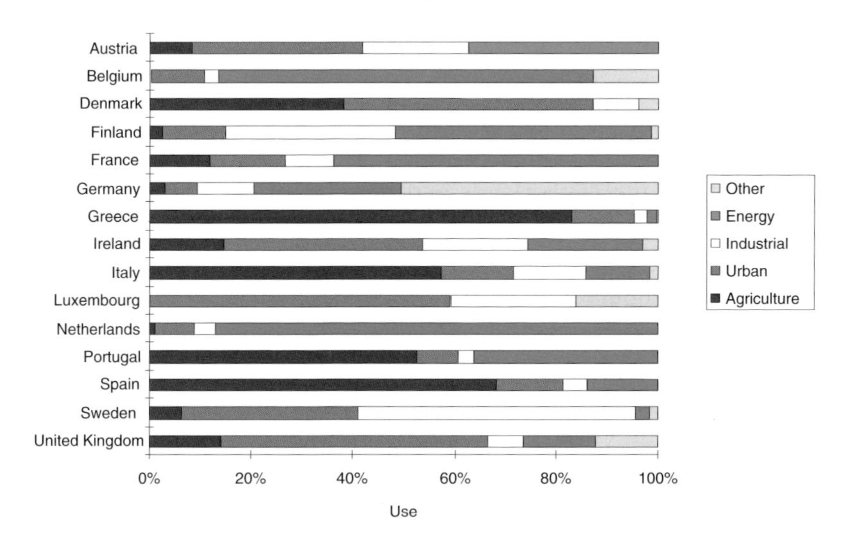

Source: Krinner et al. *(52)*.

Box 3.1. Uncontrolled irrigation in the Aral Sea region

Irrigation is the main consumer of water resources in the Amu Darya basin, which occupies a wide delta region from the Tyamyun gorge to the Aral Sea – one of the largest inland bodies of water in the world. From 1950 to 1985, the irrigated area in the basin increased four times and water intake from rivers increased by more than three times.

In the lower reaches of the Amu Darya basin, water is lost by abstraction for irrigation and by evaporation, transpiration and infiltration. Between 1950 and 1988, grain crops were largely replaced by more water-intensive crops such as rice, with a consequent increase in water used for irrigation that cannot be returned to the river. The main water pollution source in the Amu Darya basin is the return drainage water from irrigated land, as well as industrial and municipal wastewater. The inclusion of saline land in irrigated farming led to the recurrent return of salty irrigation water. The anthropogenic impact on the Aral Sea led to additional amounts of salt of natural origin from the dry sea bed entering the atmosphere, some of which also reaches the Amu Darya basin.

By the early 1990s, the mean salinity had increased to 30 parts per thousand and the sedimentation of sulfate salts had begun. At present 45% of the area of the Aral Sea has dried out, and the volume has decreased by almost 70%. The intensive salinization and high concentrations of pesticides has led to the water in the lower parts of the Rivers Syr Darya and Amu Darya becoming unsuitable for drinking. This problem is again exacerbated by the diversion of water for irrigation.

Source: Federov et al. *(53)*.

POPULATION GROWTH AND URBANIZATION

Changes in population, population distribution and density are key factors influencing the demand for water resources. The population of the EU currently exceeds 375 million, with positive growth rates in nearly all countries. The current trends, however, are not entirely clear. One long-range forecast based on Bulgaria, France, Greece, Hungary, Italy, the Netherlands and the United Kingdom predicts a decrease in population for the next three decades. Other projections show that the population is expected to increase for the next 15 years, with the total population in the current EU countries reaching about 390 million by 2010 *(54)*.

The degree of urbanization varies greatly between countries in the European Region (Fig. 3.3 and 3.4). In Belgium, Iceland, Luxembourg and Malta more than 90% of the population live in urban areas, whereas in Portugal and Tajikistan fewer than 40% do so (Fig. 3.3).

Fig. 3.3. Percentage of population living in urban areas in the WHO European Region, around 1995

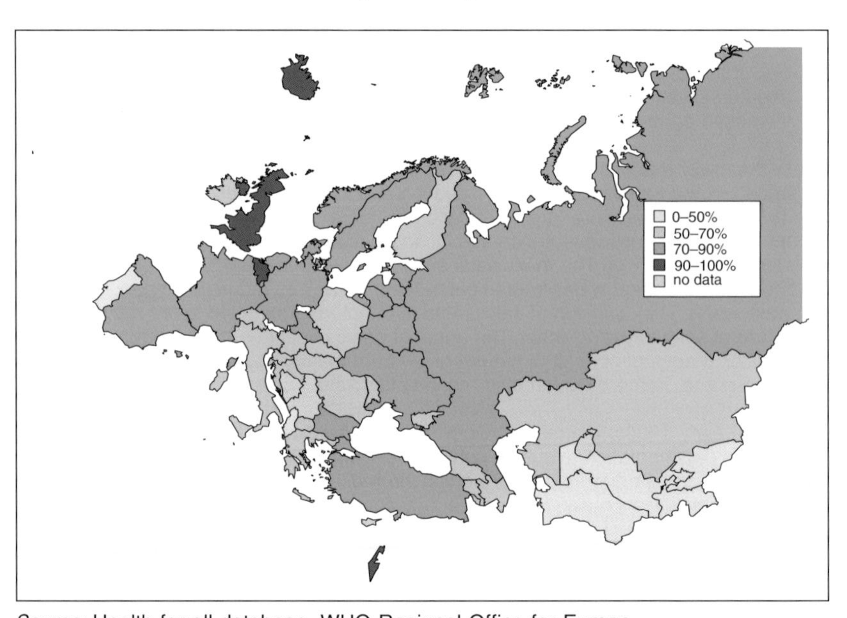

Source: Health for all database, WHO Regional Office for Europe.

Fig. 3.4. Percentage of population living in rural areas in the WHO European Region, around 1995

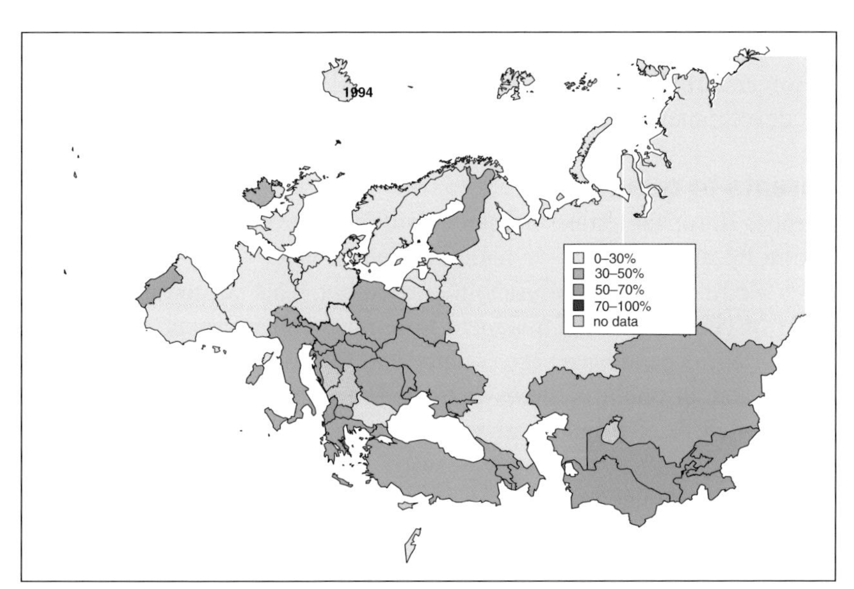

Source: Health for all database, WHO Regional Office for Europe.

Demands on quantity

The water required for drinking and other domestic purposes is a significant proportion of the total water demand. This proportion varies between countries, and comparison is not easy since the data available usually refer to public water supplies, which include some industrial use but exclude private supplies used to supply individual households or small groups. The proportion of water for urban use in total abstraction ranges from about 6.5% in Germany to more than 50% in the United Kingdom.

Increased urbanization concentrates water demand and can lead to the over-exploitation of local water resources. One consequence of increased urbanization is a change in run-off patterns resulting from large areas being covered with an impermeable surface such as concrete, tarmac or roofs. Most rainfall in cities enters a storm-drain system and is discharged, either directly or via a wastewater treatment plant, into surface waters. Thus, rainfall that in a rural area might have served to replenish groundwater supplies is, instead, directed to surface sources.

Urbanization often, though not always, accompanies increased industrialization and economic activity. The resulting rise in the standard of living is generally associated with increased water demand, for example from the use of water-consuming household appliances. However, urban water demand is expected to stabilize as a result of the development and use of appliances that are more water-efficient.

Threats to quality

In most European cities, human waste is removed from the house or latrine by a water-flush system. This wastewater enters a network of pipes along with other household wastewater. This wastewater may receive varying degrees of treatment to remove contaminants, or no treatment at all, depending on the country and the location before discharge. Wastewater in public systems is treated at municipal treatment plants, which often receive industrial wastewater as well as domestic sewage.

Although the concentration of large numbers of people in close proximity simplifies the processes of water supply and wastewater collection, disposal of the large amounts of waste generated can compromise the water quality in the recipient body of water. The release of untreated or partly treated wastewater to surface water produces contamination by microorganisms and nutrients and increases the BOD (see Chapter 2). The conventional mechanical (primary) and biological (secondary) wastewater treatment methods do not remove microorganisms or nutrients such as nitrogen and phosphorus. Additional (tertiary) treatment to remove phosphorus has been common in Finland and Sweden since the mid-1970s and is becoming more common in other western and northern European countries. In much of Europe, however, effluents discharged from wastewater treatment plants contain nutrients that contribute to eutrophication (see page 30).

Wastewater effluent is the most significant contributor of phosphorus to surface water, and detergent adds significantly to the phosphorus content of domestic sewage. Bans or voluntary agreements have been successful in reducing the use of phosphate-based detergents in a number of European countries.

Evidence also suggests that domestic sewage is a source of endocrine-disrupting chemicals in the aquatic environment. Studies of sewage discharges to British rivers have demonstrated the presence of the natural and synthetic hormones used in the contraceptive pill (see page 72) *(55)*.

Poor personal hygiene, resulting in transfer of microorganisms directly from person to person or via contaminated food, is often a significant factor in the spread of disease. This route of infection becomes more likely where water supplies are inadequate or interrupted and frequent washing is impractical. Poor sanitation and sewage disposal are often associated with the spread of enteric diseases. In such circumstances faeces can contaminate the water sources that are used for potable supply, with the consequent spread of disease.

In some areas, especially sparsely populated rural areas, installing a public sewerage network may involve excessive cost. In such cases, independent sewerage facilities such as sealed septic tanks are an acceptable alternative. Other on-site sanitation systems such as pit latrines may protect the health of the local community by improving community hygiene. If these sanitation systems are not properly designed or not adapted to the local conditions, however, pathogens may be released and contaminate local bodies of water or the environment. Fig. 3.5, 3.6 and 3.7 show the proportion of the European population connected to hygienic sewage disposal.

Even if wastewater is collected by a public sewerage system, it may receive little or no treatment before being discharged into surface waters (Table 3.1 and Box 3.2). Advanced methods of wastewater treatment are available and are used in some countries in western and northern Europe. In some of these countries, most sewage collected receives at least secondary treatment, although there are exceptions *(11)*. Some countries in southern Europe and in the eastern part of the European Region discharge significant proportions of collected sewage untreated.

Wastewater treatment, as developed and practised in large parts of western Europe, is not designed to remove pathogens efficiently. Its primary purpose is to remove solids and reduce the BOD of the wastewater so as to prevent deoxygenation and contamination of rivers when the effluent is discharged to water bodies. Nevertheless, 90–99% of microorganisms may be removed from the water phase

Fig. 3.5. Percentage of total population with hygienic sewage disposal in the WHO European Region, early 1990s

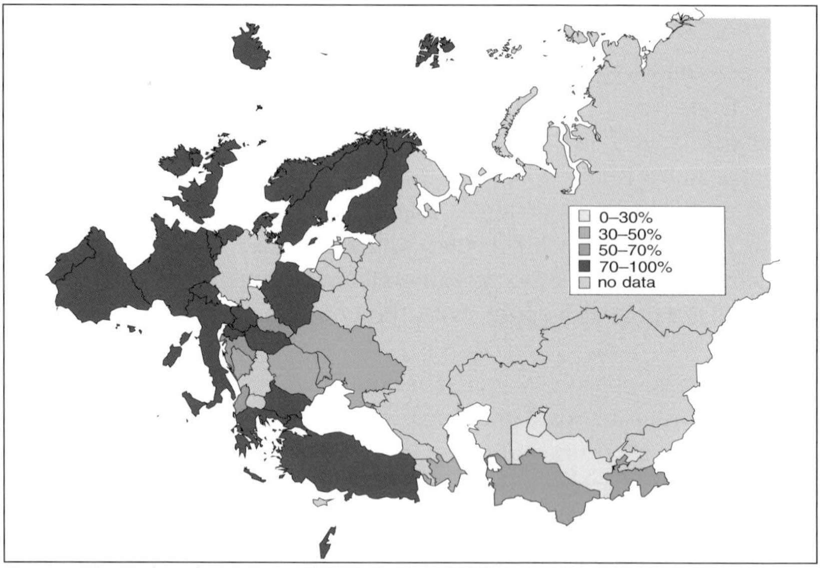

0–30%
30–50%
50–70%
70–100%
no data

Source: Health for all database, WHO Regional Office for Europe.

Fig. 3.6. Percentage of urban population with hygienic sewage disposal in the WHO European Region, early 1990s

40–70%
70–80%
80–90%
90–100%
no data

Source: Health for all database, WHO Regional Office for Europe.

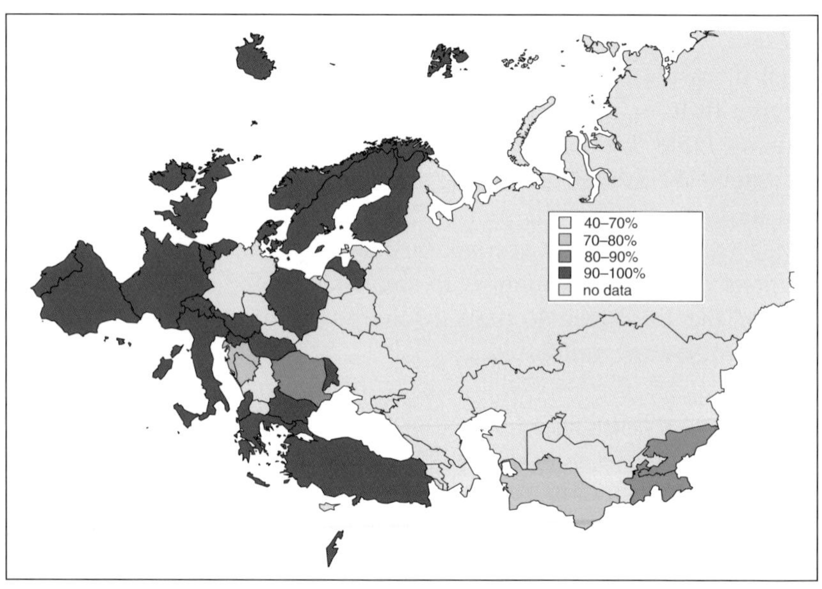

Fig. 3.7. Percentage of rural population with hygienic sewage disposal in the WHO European Region, early 1990s

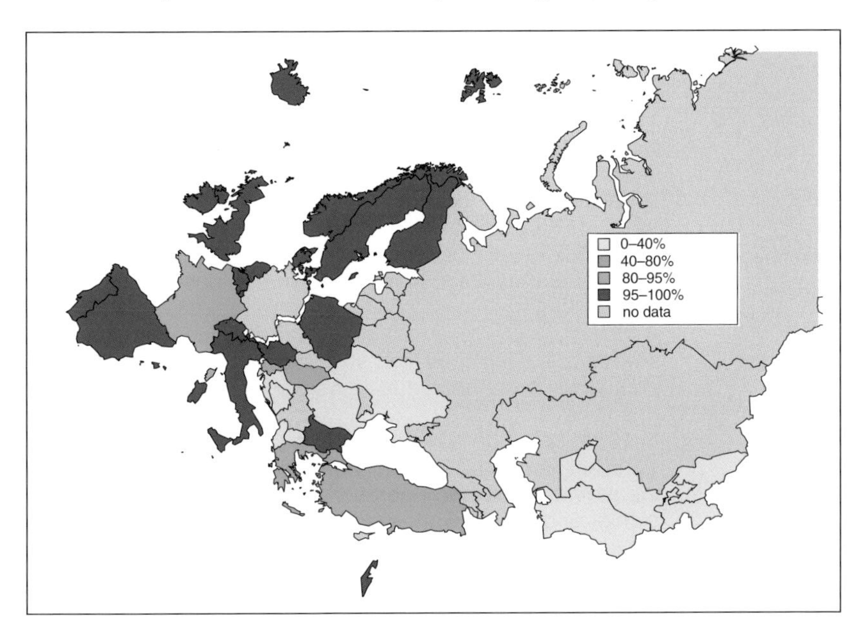

Source: Health for all database, WHO Regional Office for Europe.

during primary and secondary treatment and concentrated in the sludge. Thus, although the microbial content of effluent is lower than in the raw sewage, discharges from even quite advanced treatment plants are far from microbially pure. Life stages that do not readily settle, such as protozoan cysts, are also not well removed by conventional wastewater treatment (see Chapter 6).

As the demand for water increases, wastewater reclamation and reuse may play an important role in water resources management by providing a means to produce quality source water for irrigation and for industrial and urban requirements. In arid regions such as Israel, water reuse provides essential water for agricultural production. Where humans are potentially exposed to untreated reclaimed wastewater, the major acute health risks are associated with exposure to pathogens, including bacteria and nematodes. Treatment of wastewater is a highly effective method of safeguarding public health *(57)*.

Table 3.1. Sewage collection and treatment in selected countries in Europe

Country	Comments
Albania	There are no wastewater treatment facilities to serve the wastewater collection systems in any of the municipalities. Collected wastewater is discharged, untreated, to inland surface waters or directly to the sea.
Bulgaria	Of the 237 towns, 187 (79%) have a sewerage network. Only 5% (92) of the villages have a sewerage network.
Czech Republic	A total of 615.6 million m^3 of sewage is discharged to sewerage systems. Of this, 90.3% is treated; 70% of the population is connected to sewerage systems.
Estonia	In 1994, about 65% of small wastewater treatment plants were not working. The poor rural population cannot pay for the maintenance and operation of wastewater treatment plants, and the network of institutions providing technical servicing of sewerage systems and wastewater treatment plants has disintegrated.
Romania	Sewerage systems are found in 99.6% of urban localities and 2.6% of rural settlements. No adequate wastewater treatment plants have been developed in the localities.
Russian Federation	About 77% of the collected sewage is treated; 72% receives secondary treatment, and 10% undergoes tertiary treatment.
Ukraine	Approximately 90% of collected wastewater is treated, but parts of the treatment plants are often out of use because of problems with maintenance and lack of spare parts. There is no tertiary treatment. There are approximately 150 000 ha of sludge lagoons containing sludge that is too heavily contaminated with heavy metals to be spread on agricultural land.

Source: Mountain Unlimited *(50)*.

Agriculture and Food Production

A considerable proportion of Europe's surface area is devoted to food production. Of the total agricultural land area in the USSR in 1989, 37.3% lay in the RSFSR, 36.9% in Kazakhstan, 7.4% in Ukraine and the Republic of Moldova and 2.8% in Belarus and the Baltic republics *(7)*.

Box 3.2. Wastewater disposal in Slovenia

Most drinking-water in Slovenia is derived from groundwater. One of the most important sources is the Karats, an area of permeable limestone with many underground caves and streams that covers 44% of Slovenian territory. Because underground water flows very rapidly in this region, contamination can spread quickly and affect water sources some distance from the origin. Many surface waters in Slovenia are contaminated.

Only 44% of the Slovenian population is connected to the public sewerage system, and only 36% of the wastewater in the public sewerage system is treated before it is discharged into streams. Much of this water enters the Karats underground caves and waters. The lack of wastewater treatment extends even to the largest towns, including Mariner and the capital Ljubljana, despite a comprehensive and expensive plan implemented several years ago under which over 100 treatment plants were constructed. Unfortunately, the plants built incorporated inappropriate technology or were poorly maintained and, in general, only the primary mechanical treatment stage worked effectively.

Sewage from domestic sources, from public institutions such as schools and hospitals and from industry and agriculture causes significant pollution in Slovenia, with intensive pig farms being notable point-source polluters. An estimated 5000 wastewater treatment plants need to be constructed. Some of these would be very small, and it has been suggested that reed-bed treatments might be an effective and less expensive alternative for small towns and villages.

Source: Pucelj *(56)*.

The proportion of agricultural land that is irrigated tends to be higher in southern European countries, although this is not exclusively the case: a large proportion of agriculture in Denmark and the Netherlands uses irrigation. About 2% of the land area in western Europe as a whole is irrigated *(1)*.

Demands on quantity

The development and intensification of agriculture in recent decades has implications for both the quantity and quality of water sources. Agricultural demand for water in Europe is dominated by its use for irrigation. Thus in recent decades the trend in agricultural water use has, in general, been upwards, although the rate of increase in irrigated areas seems to have diminished more recently in several countries. From 1990, the irrigated area tended to be stable in Austria, Denmark, Finland, Ireland, the Netherlands and Portugal; the area has been decreasing in Germany, Italy and the United Kingdom, but has shown growth in France, Greece and Spain.

Freshwater aquaculture makes fewer demands on the quantity of water but can, if not carefully regulated and managed, severely affect water and ecological quality. Unlike agriculture, where much of the irrigation water is lost to evaporation and transpiration or moved between compartments, water abstracted for freshwater aquaculture is returned to the watercourse shortly after abstraction, with very little loss in quantity *(58)*. Some forms of freshwater aquaculture (caged fish farming) are undertaken in lakes and estuaries, where there is no need for abstraction, but a deterioration in quality may become apparent if the quantity of fish produced is not regulated.

The global aquaculture production statistics reported for 1984–1996 are noticeably higher than those between 1984 and 1995, owing to underreporting of Chinese production figures. By 1996, Europe was the second largest contributor, with 4.7% of world production. The transition to a market economy adversely affected aquaculture development in the republics of the former USSR, with total production declining at an average of 21% per year in all countries between 1990 and 1996. Between 1990 and 1996 Armenia reported a 13% decline in aquaculture output and Tajikistan 47%. Aquaculture production in the newly independent states is dominated by the Russian Federation and Ukraine, which accounted for 83% of total production in 1996 *(59)*.

Threats to quality

Concerns about the impact of agriculture on the quality of water resources are often related to the leaching and run-off of agricultural chemicals applied to crops and soil. Some agricultural contamination can originate from point sources, but most stems from diffuse sources. The use of agricultural chemicals depends on the type of agriculture practised within a country and the market price of the crops grown. The economic conditions of the country and the agricultural subsidies available to farmers also strongly influence the extent of use. Contamination of water by nutrients and microbial pathogens from farm waste and animal slurry are also concerns.

Microorganisms

Farmyard waste or run-off from fields on which animal manure has been spread can contain microorganisms that are human pathogens. These include the bacterium *Escherichia coli,* some strains of which are enteropathogenic, and the protozoan *Cryptosporidium*. Diseases caused by both these pathogens, originating in agriculture, have broken out in the United Kingdom recently, with significant numbers of people affected. *Cryptosporidium* outbreaks have occurred following contamination of water sources used for abstraction for potable water. Diarrhoeal diseases caused by *E. coli* strain 0157, resulting in death in some cases, have been linked to consumption of infected meat or contact with contaminated mud.

Agricultural land has been used as a recipient for sludge produced by wastewater treatment for many years. The practice is increasing in some countries as disposal to the marine environment is being reduced, and was prohibited in the EU as of 31 December 1998 under the Urban Wastewater Treatment Directive *(60)*. Domestic wastewater has also been used to irrigate crops in parts of the world that historically have fewer water resources than Europe. As well as providing water for irrigation, this also supplies nutrients to the crop and reduces the release of nutrients to surface waters in effluent. Within Europe, Germany and the United Kingdom have practised wastewater reuse, and its use has been increasing in some countries.

Nutrients

One of the major threats to the quality of surface waters in Europe is eutrophication as a result of excessive nutrient loading (see Chapter 2). The intensification of agriculture in many areas has resulted in large quantities of inorganic fertilizer being applied to arable land, resulting in eutrophication of lakes, rivers and coastal waters *(61)*. Intensification of livestock farming practices has also resulted in increased production of livestock waste, which is used as organic fertilizer *(62)*.

The application of organic fertilizers, such as animal manure, to arable land and run-off of slurry from farmyards may be another source of pollution by nutrients. Animal feeds are often high in nutrients to encourage rapid growth, and the excess is excreted.

Common inorganic fertilizer formulations contain nitrogen, phosphorus and potassium in varying proportions. Use of inorganic fertilizers varies between countries, depending on the economic situation and predominant agricultural practices (Fig. 3.8).

Fig. 3.8. Application of commercial nitrate fertilizer in selected European countries, 1994

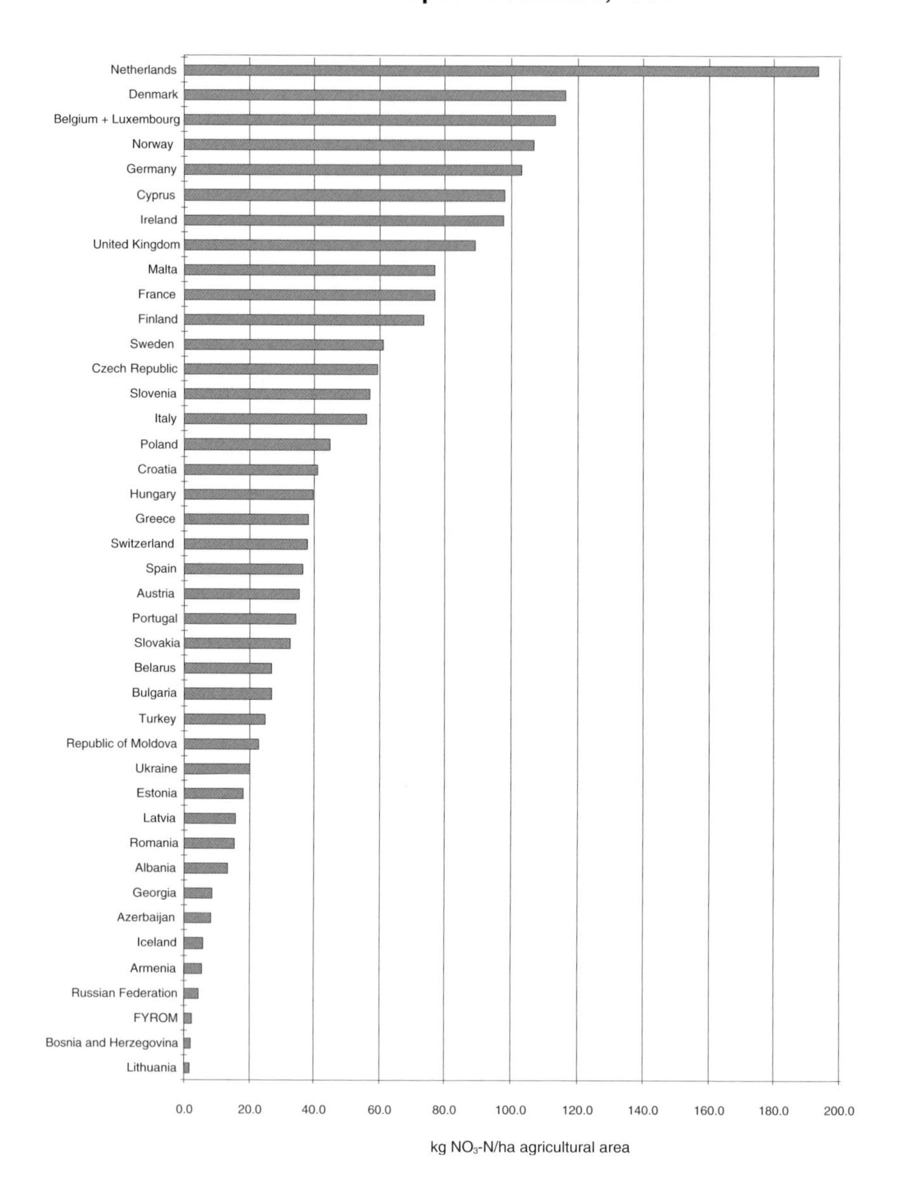

Note: FYROM: The former Yugoslav Republic of Macedonia.

Source: Food and Agriculture Organization of the United Nations *(63)*.

Although potassium applied in this manner rarely causes problems, in extreme cases it can reach very high levels in surface water. For example, the effects of intensive horticultural practices on Guernsey (Channel Islands) mean that it is only by blending water from several different sources that the standard for potassium of the EU Drinking Water Directive *(47,48)* can be met.

The system of subsidies operating in the EU under the common agricultural policy has encouraged the maximization of crop yields and has often resulted in the application of fertilizer at more than optimal rates. Nitrogen not taken up by the growing crop is then liable to leach into groundwater or run off into surface waters, depending on such factors as the precipitation pattern. This difference between the input of nitrogen to land and the output in crops (the nitrogen balance) largely determines the potential for nitrogen leaching. Nitrogen balance studies in EU countries have shown that the nitrogen surplus is as high as 200 kg/ha per year in the Netherlands, but less than 10 kg/ha per year in Portugal. This demonstrates the range and influence of different farming methods, even within the EU subsidy system (Fig. 3.9).

Although point sources of pollution, such as sewage discharges, can contribute significantly to nitrogen pollution in some regions, diffuse sources such as agriculture are usually the major contributors. In some countries, such as the Netherlands, the proportion of water pollution caused by diffuse sources is steadily increasing. Nitrogen is usually the limiting factor for the growth of algae in marine waters, and leaching of nitrate fertilizers causes marine eutrophication. Nitrogen can also limit the rate of growth in fresh surface waters at certain times, and may therefore also contribute to algal blooms in surface waters. Nitrate readily leaches into groundwater, where it may cause health problems if water is abstracted for potable supply *(62)* (see Chapter 6).

Agriculture also contributes to phosphorus loading of surface water, and hence to eutrophication and associated ecological problems. As with nitrogen, intensive farming methods produce a surplus of phosphorus in the soil, which has been estimated to be 13 kg/ha per year in the EU *(65)*. High surpluses have been reported in Belgium, Denmark, Germany, Luxembourg and the Netherlands. However, in

Fig. 3.9. Nitrogen balances for the soil surface of agricultural land in the 12 EU countries in 1993: countries ranked by surplus of input over output

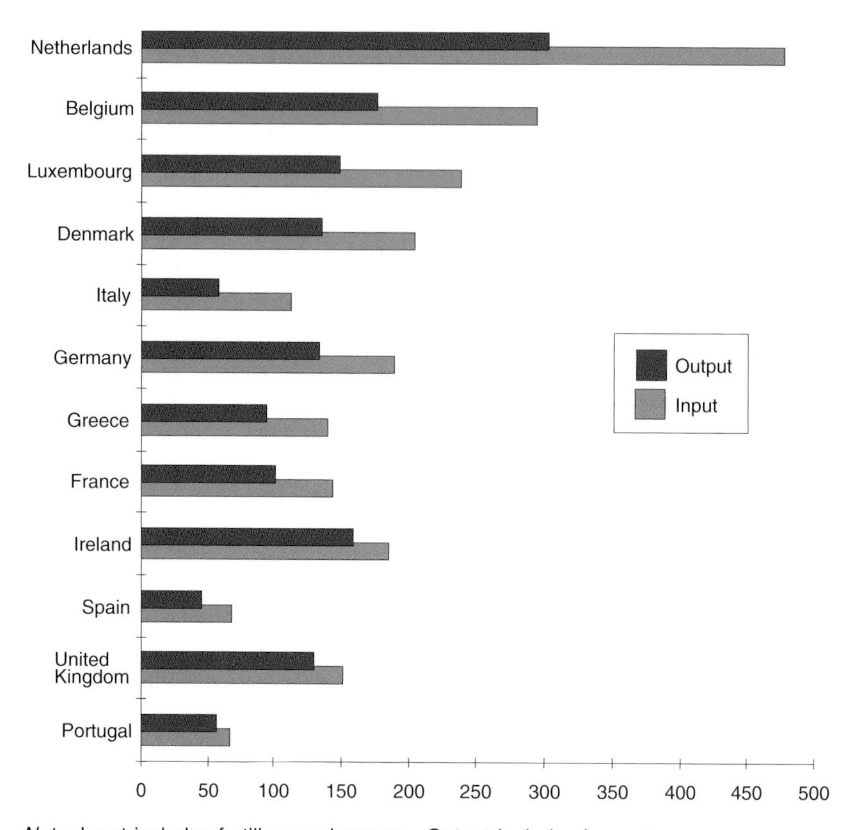

Note: Input includes fertilizer and manure. Output includes harvest.

Source: Eurostat *(64)*.

densely populated areas this contribution has largely been insignificant compared with phosphorus from point sources, predominantly sewage discharges. As the control of pollution from urban wastewater improves, phosphorus from agricultural sources may become more significant.

Aquaculture is another major source of nutrients in some lakes and rivers. Although reported nutrient export rates from fish farms can vary substantially, most values appear to be about 20–100 kg of phosphorus per tonne of salmonid fish *(66,67)*. It might be expected that the advent of fish food low in phosphorus would have reduced these export coefficients, but how widely farmers have adopted these feeds is not known.

Pesticides

The use of pesticides in agriculture has become commonplace over the last half century. In arable cropping, herbicides are used in the greatest quantities, and this type of pesticide has been detected most frequently in European groundwater (see Chapter 2). Fungicides are used to prevent plant diseases, and insecticides are applied both to prevent direct damage by insects and to prevent the spread of viral diseases for which insects are vectors. Modern animal husbandry also involves the use of veterinary products such as insecticides for use in animal houses, and dips used to prevent infestation of the animals by ectoparasites.

As with fertilizers, pesticide use depends on economic conditions, the market price for the crop being grown and the subsidies available, as well as accepted practice in the country concerned. Use varies widely across Europe (Fig. 3.10).

Within the EU, a decreasing quantity of pesticide active ingredients has been sold since 1985 (Fig. 3.11). This may result, in part, from changes in the pattern of use of established pesticides, but it also reflects the increased activity of recently introduced pesticides, which enables them to be effective at lower application rates. Long-term monitoring of water bodies in Armenia, Azerbaijan and Georgia has shown an overall decreasing trend in concentrations of organochlorine pesticides since 1978 *(24)*.

Groundwater is usually regarded as being most vulnerable to pollution by pesticides, because of the long residence time and minimal degradation. However, pollution by pesticides can also affect surface water and can have toxic effects on aquatic life. Lowland surface water is particularly likely to contain pesticides, although some products, such as those used in sheep-dips, may find specific use in upland areas. Some pesticides, such as the sheep-dip chemical cypermethrin, bind very strongly to sediment, and this can act as a reservoir that will slowly release the chemical over a long period.

Fig. 3.10. Pesticide consumption in kilograms per hectare of arable land and permanent crop land in selected European countries

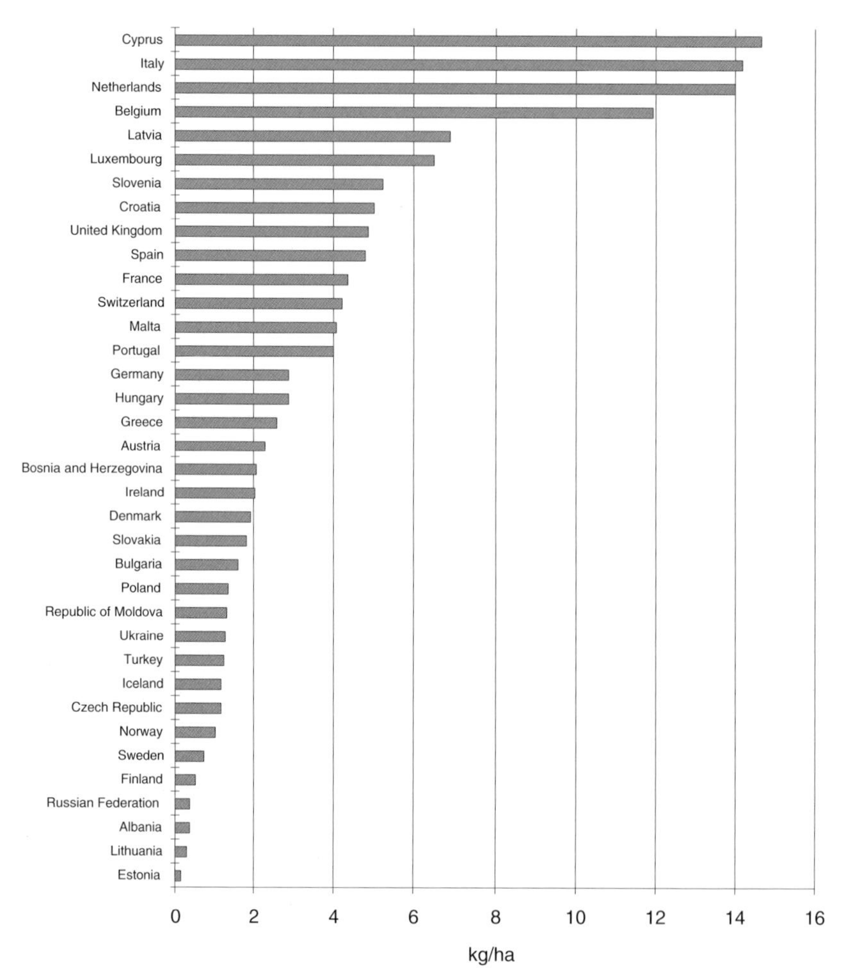

Source: Scheidleder et al. *(32)*.

As well as run-off or leaching from applied pesticides, pollution can result from the inappropriate disposal of excess spray, tank washings and spent sheep-dip (Box 3.3). Antibiotics, vaccines and other chemicals, such as malachite green (for the control of fungal infection), are widely used for the control of disease in fish farming. Most of these chemicals are administered as food additives, but chemicals utilized in the control of ectoparasites (such as acriflavine) are

Fig. 3.11. Total sales of pesticides in EU countries, 1985–1995

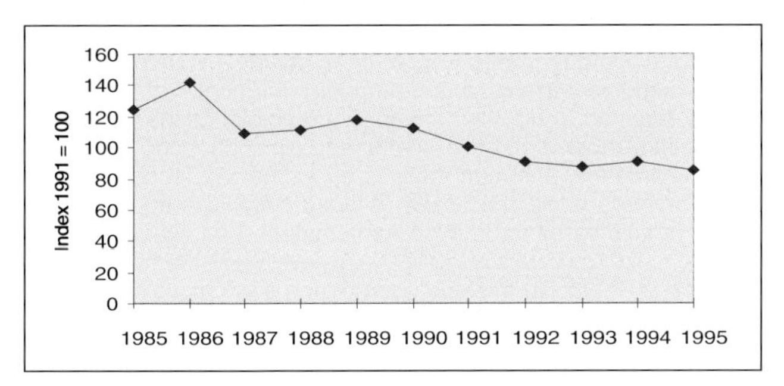

Source: European Crop Protection Association *(68)*.

often administered as dip treatments. Pesticides have been implicated in various disorders and diseases, including cancer, adverse reproductive disorders, impaired immune function and allergic reactions, especially of the skin (see Chapter 6).

Suspended solids

Soil erosion is influenced by the degree of planting and the tillage regimes used in agriculture. Run-off containing high levels of particulate matter results in contamination of source water by suspended solids, and heavy rainfall on fields without a crop root structure in place accentuates this. Thus, the type and timing of cultivation can influence water quality. For example, Lee et al. *(71,72)* found in Great Britain that the increase in lowland erosion results from the adoption of winter cereals and the consequent expansion of the area left bare in autumn and winter *(73–75)*. Other factors that are believed to have contributed to the increase include:

- arable farming on steep slopes;
- the removal of field boundaries to create larger fields *(74)*;
- an inappropriate choice of crops on steep slopes and erodible soils;
- working land up and down the line of maximum slope;
- the presence of vehicle wheel tracks, which act as channels for runoff *(76)*; and
- rolling of seedbeds *(75)*.

Fish farming is also a source of relatively large amounts of suspended solids, consisting of uneaten food pellets, faecal particles

Box 3.3. Disposal of spent sheep-dip in the United Kingdom

Until 1992, the dipping of sheep to control ectoparasites such as scab and blowfly was compulsory in the United Kingdom. Although farmers are no longer legally required to dip their sheep, good animal husbandry may require it. In recent years most dips have contained organophosphorus compounds such as propetamphos, although there is an increasing tendency, largely because of concerns about the health of those administering the dip, towards the use of synthetic pyrethroids such as flumethrin and cypermethrin.

Large volumes of spent dip are produced during each dipping operation, presenting the farmer with the problem of disposal. Older dipping tanks incorporated a drain plug, which was removed when the dipping operation was over. The preferred disposal method for many years was a "soakaway", a large pit backfilled with stones and sand or gravel. Because intensive sheep farming is largely located in upland areas with shallow soil and a high water table, however, these methods often resulted in most of the dip entering streams and rivers. On-site treatment of the dip using flocculation, settlement and filtration by gravel and granulated active carbon has been considered, but was found to be unsuitable for the disposal of spent dip because of the high solids and lanolin content *(69)*.

Disposal through an approved waste contractor is often prohibitively expensive for many sheep farmers, even if it is the preferred option. As an alternative, the United Kingdom Environment Agency *(70)* recommends spreading thinly onto grassland at an application rate of up to 5000 litres/hectare. Co-spreading with slurry is suggested, as this may increase the rate at which the active ingredients are broken down. Level ground should be used, and spreading should not be carried out close to watercourses or springs, or if rain is forecast or if the soil water is already at field capacity. The guidelines stress that the spent dip should only be spread if it will not affect the quality of groundwater or surface water. However, the concentration of sheep farming in hilly areas with thin soil and high rainfall is likely to preclude the implementation of these safeguards in many cases.

The leaching of sheep-dip into aquifers is also a potential concern, although the major aquifers in the United Kingdom are not located where sheep farming is practised most intensively. Nevertheless, sheep are farmed, for example, in the South Downs, and percolation into the saturated and unsaturated layer has been demonstrated following the repeated spreading of spent dip from a mobile operation *(69)*. Although the risk that sheep-dip disposal imposes on the quality of groundwater sources is not as great as that for surface water, high concentrations in rivers and streams are seasonal, coinciding with sheep-dipping operations. In contrast, contamination of an aquifer will lead to a more constant and persistent problem.

(including gut bacteria), fish scales, mucus and other detritus. Estimates of dry pellet loss from trout and salmon tank and pond culture are commonly in the range of 5–20%, with feed losses from cage farms thought to be greater *(66)*. The use of settlement ponds and other treatment methods, together with different sampling frequencies, system types, feeding methods and rates, fish sizes and types of diet results in a wide range of loads of suspended solids from fish farms. Values reviewed by the Nature Conservancy Council *(66)* range from 110 kg to 2153 kg per tonne of fish produced. This load of solids has an associated oxygen demand, usually accounting for at least half of the total BOD (particulate and dissolved) from salmonid fish farms. Fish farm effluent does not usually significantly affect dissolved oxygen levels in running water, but organic enrichment below caged fish farms in lakes is often detrimental to the sediment invertebrate community.

INDUSTRY AND TRANSPORT

Industrial demand and effects on water quality can directly affect the water supplies of a large number of people where industry co-exists with highly populated urban areas.

Demands on quantity

Some industries, especially traditional heavy industries, require large amounts of water for cleaning or cooling and therefore compete for water resources. The amount of water abstracted for cooling usually far exceeds that used during industrial processes, but this is often regarded as a non-consumptive use, as the water is returned to its source virtually unchanged except for an increase in temperature and, in some instances, the presence of a biocide. The amount of water required also depends on the type of process used; in general, more modern plants would be expected to incorporate more water-saving measures.

The amount of water used by industry and its significance as a proportion of total abstraction varies greatly between countries (see Fig. 3.2). Figures vary according to the different methods used to record specific uses. For example, some countries include cooling water in the figure for industrial use while others do not. Abstraction for industrial purposes in Europe has been decreasing since 1980, largely because of industrial recession in many countries in the eastern part of the Region and changes in the predominant industries and industrial practices in northern and western Europe. In the latter countries traditional industries such as textiles, iron and steel have declined, and this is coupled with technological improvements in water-using equipment and increased recycling.

Threats to quality

Industrial processes produce contaminated wastewater that is released into fresh or marine surface water either directly or following treatment. The range of chemical contamination that may be released is large, but much attention has been paid to substances that may accumulate in sediment or bioaccumulate and enter the food chain, such as certain heavy metals and organic substances.

In addition to controlled or intentional discharges, contamination can also occur as a result of spillage, poor handling, improper disposal methods and accidents. Acute pollution of source water can occur, for example, following road accidents involving chemical tankers or a fire. Continuing pollution may result from leaking oil pipelines or chemical tanks.

Pollution of water sources is often a legacy of previous industrial practices. This is especially the case for groundwater, in which contamination persists for many decades because of the lack of volatilization and degradation and the slow recharge time. Leaching of industrial chemicals into groundwater has also resulted from spillage and the historical disposal of chemicals on to land (Box 3.4).

In some cases, the discharge may continue even after the industry has ceased, such as the release of contaminated flood water from disused mines. Nevertheless, discharges will, in general, change as

Box 3.4. Contamination of groundwater at military sites

Armed forces have caused environmental damage during peacetime in many parts of the world. In Hungary, for example, environmental impact assessment studies indicated serious environmental pollution at 18 Russian military bases, including airfields. The major problems were related to the uncontrolled release of jet fuel just before landing, directly affecting the soil and groundwater at the ends of the runway. This resulted in serious contamination of the groundwater – a kerosene layer was observed floating on the top of the groundwater at several locations. Dispersion, diffusion and dissolution of components of this kerosene, especially one-ring aromatic hydrocarbons such as benzene, toluene, ethylbenzene and total xylene (BTEX) in the groundwater, and the migration by lateral flow into aquifers used for drinking-water abstraction, created potential risk to consumers. In addition to the inappropriate disposal of jet fuel, other potential environmental hazards involve the unsecured disposal of hazardous materials that are often unidentified. This may include toxic chemicals buried in the area of the military base.

the dominant industries change and industrial practices and disposal techniques improve.

Many newly independent states recognize that industrial pollution is poorly controlled, and detrimental effects on health have been reported. In the Russian Federation, for example, illnesses have been linked with drinking-water polluted by industry in the Perm'-Krasnokamsk industrial zone, the cities of Kemerovo and Yurga, and near the Ust'-Ilimsk plant *(77)*.

Heavy metals

Historically, heavy metals have been common contaminants, being released from a range of industries. Many accumulate in the body and can persist in sediment, being released into the water column when the sediment is disturbed. However, measures undertaken in the Nordic and western European countries to reduce heavy metal emissions to inland waters and marine areas have been successful.

Petrochemicals and volatile organic compounds

Petrochemicals are among the most widely used of all industrial products because of their use in transport and power generation. Pollution incidents arising from spills from tanker disasters in the marine environment have, perhaps, been the most widely publicized incidents involving this industry. However, freshwater sources have been contaminated by leakage from petrochemical plants and pipelines, military sites and fuel tanks at filling stations (Box 3.5). Groundwater is vulnerable, and contamination by hydrocarbons is a particular problem in parts of eastern Europe. In the Czech Republic, for example, approximately 50% of accidental water pollution incidents involve petroleum *(78)*. Some regions of Estonia also report contamination of surface water and groundwater by phenols and petroleum products.

Contamination of aquifers with hydrocarbons may also influence the distribution of other contaminants within the aquifer. The Luton and Dunstable aquifer in the United Kingdom, for example (Box 3.5), was found to be contaminated with hydrocarbons as well as organic solvents. The hydrocarbons formed a layer at the top of the aquifer into which the solvents were preferentially dissolved *(79)*.

Box 3.5. Industrial contamination of groundwater: the Luton and Dunstable aquifer

The first in-depth study of the contamination of groundwater sources in industrialized urban areas of Europe was carried out in Milan in the early 1980s *(80)*. Since then, investigations of several aquifers in the United Kingdom with different types of geology have repeated the findings of widespread contamination by industrial compounds, especially chlorinated solvents *(79,81–83)*.

The chalk aquifer lying underneath the conurbation of Luton and Dunstable in southern England is the most widely exploited groundwater source in the United Kingdom. Industry in the area is dominated by the automobile industry, which makes heavy use of chlorinated solvents as degreasing agents. The solvent traditionally used is trichloroethene, although this is being replaced in some cases by 1,1,1-trichloroethane. A related solvent, tetrachlororethene, has been widely used as a dry-cleaning fluid *(79)*.

Analysis of water from boreholes in the chalk aquifer showed a pattern of widespread low-level contamination by these solvents; some hot-spots have higher concentrations. Much of the contamination is believed to be the legacy of traditional industrial practices such as disposal by spreading on open ground to allow evaporation. Although disposal methods have changed, however, studies have found that other handling practices in many factories have changed little in recent decades. Casual use and indiscriminate disposal are still believed to be occurring. Solvents are frequently stored in drums in unbounded areas, and measures to prevent spills during decanting may be inadequate. Degreasing is often carried out in dip tanks, from which solvent loss by leakage may go undetected because it is indistinguishable from loss by solvent evaporation. The assertion that the contamination is unlikely to be entirely a legacy from the past appears to be supported by the fact that 1,1,1-trichloroethane, which has only been widely in use since the 1970s, has been found in the aquifer *(79,83)*.

Chlorinated solvents have high specific density and low viscosity. These properties make them highly mobile in many types of soil, resulting in the contamination of groundwater. In addition, they are immiscible with water and this, coupled with their density, may result in a liquid phase that is able to sink deep into the aquifer, producing widespread contamination. The solvents are chemically stable and are highly resistant to microbial degradation, making them persistent once they have entered groundwater.

Because elevated levels of chlorinated solvents were detected in the Luton and Dunstable area, the Lee Valley Water Company installed air-stripping water treatment for water extracted from boreholes. This has reduced solvent concentrations by 95% *(79)*.

A number of volatile chlorinated organic compounds are widely used in industry as solvents, degreasers and cleaning agents. These include compounds such as trichloroethene and tetrachlororethene. Inappropriate handling procedures and disposal techniques, such as spreading on the ground, have resulted in groundwater becoming contaminated through leaching. These compounds evaporate readily, and their release to surface water affects water quality less.

Hormonally active chemicals

Much public attention in recent years has been paid to the presence in the aquatic environment of chemicals that mimic natural hormones, together with concerns that they may be linked to reported increases in testicular and breast cancer and decreasing sperm counts. In some cases in Europe, adverse endocrine effects or reproductive toxicity in birds and mammals have coincided with high levels of anthropogenic chemicals that have been shown to have endocrine-disrupting properties *(84)*. Evidence for this association is limited, but some substances have been shown to cause feminization of male fish with the production of vitellogenin, an egg-yolk protein normally only found in sexually mature female fish, and oocytes. A wide range of industrial chemicals has been shown to trigger this activity (Box 3.6).

TOURISM

Tourism is an important source of income in certain areas – including a number of Mediterranean countries, Austria, Hungary, Ireland and Switzerland *(87)*. The seasonal influx of large numbers of people can significantly affect water resources and greatly increase the volume of wastewater requiring treatment and disposal.

Demands on quantity

Domestic water use by tourists is often twice that of residents, and large volumes of water are also required for recreational facilities such as swimming pools, water parks and golf courses. Areas popular for tourism are often warm areas, thus concentrating the demand in an area where water resources may already be limited (such as the Mediterranean coast of Spain), and demand often peaks during periods when the renewal of water resources is low. Localized shortages of water may therefore be common in areas used for tourism. Supplying sufficient water at the time of peak demand, often at the driest time of year, may require the construction of additional reservoirs.

Threats to quality

A large seasonal influx of tourists can produce challenges in the design and operation of water supply systems and facilities for

Box 3.6. Estrogenic substances and water

Concern has been expressed over the possible adverse environmental effects of hormone-like substances on humans and other animal species such as fish. Research to date has mainly focused on the potential for compounds to inadvertently mimic the biological activities of the female endogenous hormone estrogen, which can cause a feminizing effect. Exposure to such estrogenic chemicals has been postulated to cause effects that include reduced reproductive function, increases in certain types of human cancer, and declining wildlife populations.

Evidence exists that wastewater effluent can contain substances causing estrogenic effects, generally been demonstrated by the production of vitellogenin in fish. Vitellogenin production appears to be a very sensitive biomarker of estrogenic activity, although a clear direct relationship between its production and effects on fertility has not been established. The extent to which these observations are associated with significant changes in population viability remains unclear.

Numerous compounds have been shown to have weak estrogenic activity, such as alkylphenols, bisphenol A, polychlorinated biphenyls and some pesticides. However, natural hormones appear to be predominantly responsible for the estrogenic activity in domestic effluent *(55)*. For wastewater treatment plants that receive significant industrial input, some effluent can contain relatively high concentrations of other weak estrogenically active compounds such as alkyl phenols and alkylphenol ethoxylates. In such cases, these latter compounds are likely to contribute to the estrogenic activity, which could give rise to localized environmental impact.

Since river water is widely used for drinking-water production, this has also raised concerns that drinking-water might be a source of exposure to estrogenic substances. However, no estrogenic activity was detected at drinking-water intakes or storage reservoirs *(85)*, and analyses using methods with appropriate limits of detection have provided no evidence for the presence of free natural hormones in drinking-water obtained from rivers that receive wastewater effluent *(86)*.

wastewater collection and treatment. Large variations in the quantity of wastewater to be treated makes designing and operating efficient sewerage systems and wastewater treatment plants difficult.

Some areas that attract tourism are in terrain that makes connection to conventional sewage collection and treatment systems difficult or inappropriate. For example, the Austrian and German Alpine Associations have more than 750 refuges and lodges in the Austrian Alps, attracting about 1 million overnight stays per year. This is in addition to the 1.5 million day visitors per year, 2000 full-time staff and the 300 lodges belonging to other organizations. Tourism in the Austrian Alps has been estimated to generate a total wastewater load equivalent to 430 000 people during the holiday season. This produces technical and financial difficulties associated with treating fluctuating discharges at low temperatures *(1)*.

In summer, mass tourism concentrates along coastal areas. The increase in visitors to coastal areas at specific times of the year creates demanding health problems. Discharged urban storm water and wastewater are the main source of sea pollution that affects the quality of coastal seawater and can make it unsuitable for swimming and a threat to human health. If municipal sewage constitutes a significant source of phosphorus pollution, phosphorus must be removed at treatment plants. Heavy seasonal tourism increases the demand on the capacity of treatment plants and sewerage and also causes substantial fluctuation in the sewage load. Sewerage and treatment for fluctuating amounts of sewage present specific technical difficulties. Lake Balaton is an example of such a situation, where the number of tourists during July and August is twice that of the local population *(88)*.

If domestic wastewater is used in agriculture, health risks should be avoided by following the WHO guidelines for the use of wastewater in agriculture and aquaculture *(89)*. Areas with low population that are affected by tourism may need special consideration because the population may increase temporarily by several times and overload wastewater treatment capacities. In temperate regions, the tourism season may coincide with the cyanobacterial growth season *(37)*.

4

Access to safe water

A reliable supply of clean drinking-water is essential to protect the health of individuals and communities. Both the quantity and the quality of supply are important. An adequate quantity of water is of primary importance in public health, since diseases are more easily transferred directly from person to person or via contaminated food when poor hygienic practices occur because of insufficient water. The potential consequences of microbial pollution are such that control of drinking-water quality must never be compromised. A number of serious diseases can be spread via contaminated drinking-water, such as cholera and typhoid fever, as well as common enteric diseases such as gastroenteritis. A supply containing high levels of chemical contaminants may also significantly affect the health of a whole community.

Coverage of the drinking-water supply

A reliable and adequate source of clean drinking-water is considered to be a basic human right and is one of the highest priorities of any country. The way in which people obtain their water depends on the natural and financial resources of a country and historical influences. The population density and pattern of habitation also influence the extent to which consumers are supplied by piped networks or rely on local sources for drinking-water.

The residents of the towns and cities of Europe are generally well supplied with running water (Fig. 4.1–4.3, Table 4.1). Installing a

Fig. 4.1. Percentage of total population with home connection to water in the WHO European Region, around 1990

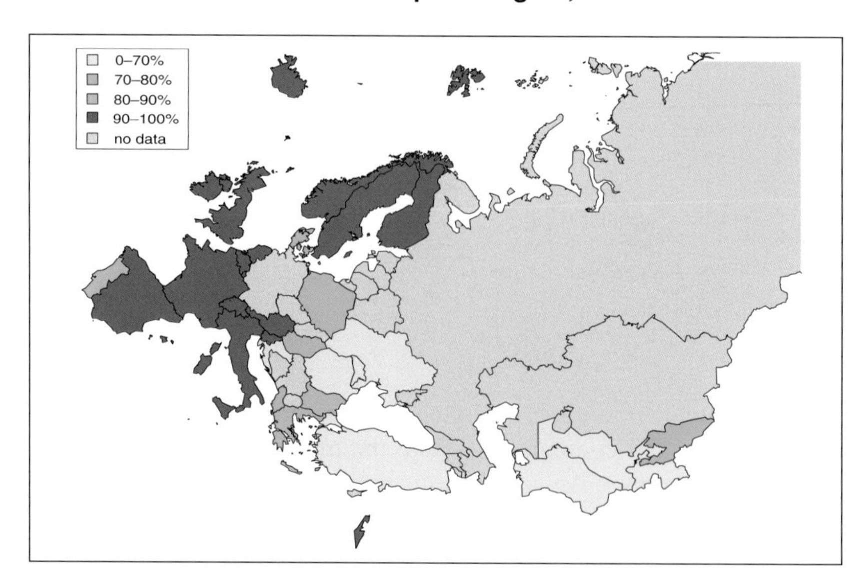

Source: Health for all database, WHO Regional Office for Europe.

Fig. 4.2. Percentage of urban population with home connection to water in the WHO European Region, around 1990

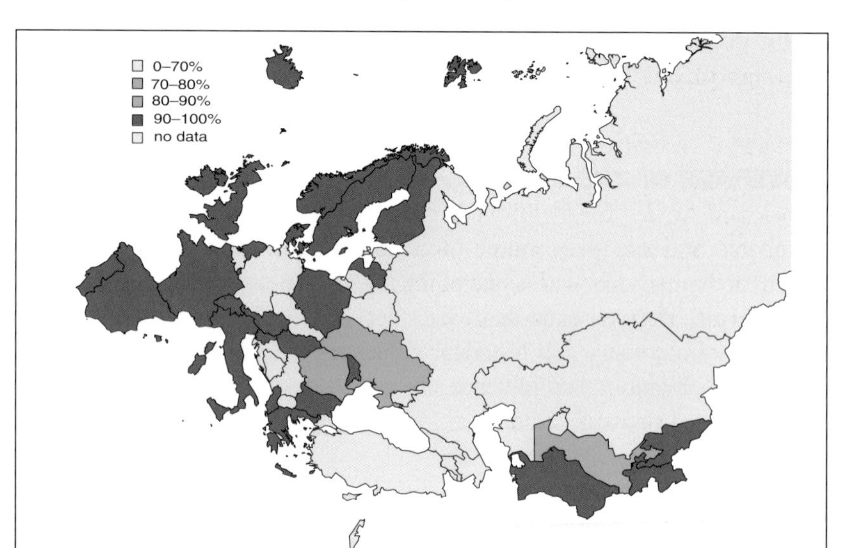

Source: Health for all database, WHO Regional Office for Europe.

Fig. 4.3. Percentage of rural population with home connection to water in the WHO European Region, around 1990

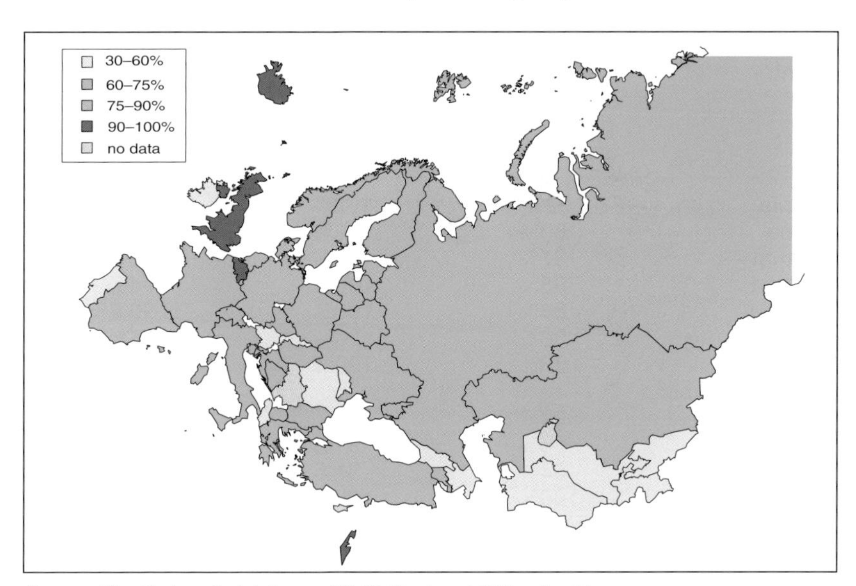

Source: Health for all database, WHO Regional Office for Europe.

water network is a large capital project and, once it is installed, maintenance is required to ensure its continued efficient operation (Box 4.1). Financial restrictions may prevent the installation of a distribution system or result in the deterioration of a network already in place.

Because of logistical difficulties, political priorities and relative cost, rural populations are less likely than urban populations to have piped water and house connections. However, the historical and current economic and organizational status of a country also strongly influences the extent of water infrastructure. For example, similar proportions of the populations of Croatia and Finland are rural (about 35%), but 87% of the total population of Finland is connected to a public water supply (1996 data) compared to only 70% of the population of Croatia (1997 data).

In sparsely populated areas of some countries, providing piped water may not be economically viable, and rural populations are more likely to rely on small, private non-piped supplies. The proportion of the population connected to public water supplies can vary

Table 4.1. Percentages of the population served by piped water supply in selected countries of the European Region

Country	Total population served by piped public water supply	Urban population served by house connection	Rural population served by piped water supply at home
Albania	92%	100%	88%
Austria	80%	100%	70%
Belgium	97%	100%	90%
Bulgaria	98%	100%	94%
Denmark	88%	100%	99.98%
Finland	80%	96%	85%
France	98%	100%	95%
Federal Republic of Germany	97.8%	100%	97%
Greece	86%	91%	73%
Hungary	84%	91.5%	74.3%
Iceland	100%	100%	100%
Ireland	90.6%	98.7%	80.5%
Israel	99.9%	100%	98%
Italy	98.8%	100%	96%
Luxembourg	99%	100%	97.6%
Malta	98%	100%	96%
Monaco	100%	100%	–
Netherlands	99.2%	99.8%	95%
Norway	87%	100%	34.7% (public) 65.3% (private)
Poland	79.9%	93.1%	55.8%
Portugal	58%	97%	50%
Romania	52.3%	91%	17%
Spain	80%	90%	50%
Sweden	86%	100%	18%
Switzerland	99%	100%	99%
Turkey	69%	72.8%	66%
USSR	–	98%	86%
United Kingdom	99%	99.5%	91.5%

Source: WHO Regional Office for Europe *(90)*.

significantly between different areas of the same country. For example, 78% of the population in the north-eastern part of Italy is connected to a public supply, compared with only 27% of the population of the Italian islands *(91)*. In some countries with a history of good water supply infrastructure, all the rural population is connected to a

Box 4.1. Water distribution systems

Distribution of treated water in large piped networked supplies requires an extensive and elaborate system of pumping stations, service reservoirs and pipework. The raw water resources and the treatment plants are often some distance from the urban populations. Reservoirs of treated water within the distribution network allow treated water to be stored close to consumers, ensuring that supply is sufficient to meet peak demand. Reservoirs can also be used to mix water from different sources, to compensate for any variation in quality by diluting the contaminants. These reservoirs are usually constructed of concrete, although smaller ones may be made of steel. They must be watertight, both to prevent leakage and to prevent contamination of the stored water by leaching from the surrounding soil.

Trunk mains carry large volumes of water, often over long distances. The smaller distribution mains take water from the branched network of pipes supplying water to the individual houses. The distribution network usually consists of a ring main (a loop) from which the service pipes to the individual houses are supplied. However, in some cases a dead-end spur has to be installed because of the housing pattern. Water can sometimes remain in a long spur for considerable periods of time before it is used, and this may adversely affect the water quality.

home water supply. This is the case, for example, in Iceland and Norway. In contrast, the homes of as few as 5% and 12% of the rural populations of Turkmenistan and Ukraine, respectively, are connected to a water supply. In Romania, 84% of the urban population is supplied by the centralized system versus 32% of the rural population. The dichotomy between provision for urban and rural populations is perhaps best illustrated by the situation in the Republic of Moldova: 98% of urban inhabitants have home connections versus only 18% of those in rural areas (Fig. 4.2 and 4.3).

Community-managed (private) supplies are usually wells or boreholes supplying local residents with groundwater. However, the wells may be very shallow and therefore prone to contamination from the surrounding agricultural land and from excreta. In some countries, water supplies from shallow wells close to surface waters are commonly used. This rudimentary bank filtration may also be prone to contamination.

Because of the capital costs involved in water treatment, small community-managed supplies, especially those supplying only a single household, often do not receive adequate treatment. More than 20%

of the population of Slovakia uses drinking-water from domestic wells, and an estimated 80–85% of such wells do not comply with national drinking-water standards *(92)*. Nitrate and phosphate concentrations in almost 17% of Latvian wells exceed 50 mg/l, especially the shallow ones. This is primarily the result of poor abstraction management and poor construction.

Continuity of drinking-water

Access to a supply of drinking-water goes beyond the presence of a well or a connection to a supply network. Most public water supplies within the EU and other countries in the western part of the Region maintain a continuous supply of water. A number of countries in central and eastern Europe, including some newly independent states, also provide continuous public water supplies. Nevertheless, some areas in a number of European countries do not receive a continuous supply of water. This may be for reasons such as shortage of source water (which may be seasonal), demand exceeding the capacity of the source or the supply system, accidents and emergencies, leakage and misuse. Discontinuity in supply may have implications for human health comparable to those where there is inadequate water. This is made worse if the disruption is unpredictable or unannounced. Discontinuity is often caused by the poor design and condition of the waterworks, and inadequate operation, maintenance and management. Financial constraints may also be responsible for interrupted supply and may be linked to a discontinuous electricity supply that prevents the continuous pumping and treatment of water (Table 4.2). In Latvia, poor infrastructure, no regular maintenance and financial constraints on collective farms that maintain the water pipes and on the responsible municipalities have resulted in an insufficient domestic water supply for rural populations.

The quality of the water being supplied may also be affected by discontinuous supply, as contaminants will enter through leaks in the network when the pressure drops. This is clearly illustrated in Armenia, where faults in the network when the supply is switched off have been associated with contamination and outbreaks of waterborne disease (Box 4.2) *(93)*. In the Russian Federation, about 50% of the population uses water that does not meet national quality

Table 4.2. Countries in the European Region reporting discontinuity of drinking-water supply

Country	Problems in achieving a continuous supply of drinking-water
Albania	Supplies are provided intermittently for a number of hours each day (1–3 times per day or 1–3 hours per day). Interruptions occur all year round, and 100% of the population is affected. Poor management, low levels of maintenance, limited funds for repairing defects, poor availability of equipment and increasing demand contribute to discontinuity. Providing sufficient pressure is often impossible.
Iceland	There are no major problems, but some regional problems are possible.
Italy	An estimated 18% of households suffer from persistent discontinuities in the water supply. This varies from 8% in the northeast to 30% in the islands.
Latvia	The electricity supply is discontinuous (during emergencies) and equipment is not readily available.
Malta	Although all urban areas are provided with a continuous supply (except during power failures), some new building developments have to use water tankers as they have not yet been connected to the drinking-water network.
Republic of Moldova	Supply interruptions are very frequent, especially in rural areas, villages and small towns. About 75% of the population is affected. Problems include a discontinuous electricity supply, water shortages, financial considerations and poor availability of equipment.
Romania	Problems include urban development without adequate facilities, financial considerations, deficiencies in network systems and low capacity for storage. In 1993, 37% of the total population connected to a piped water system received interrupted supplies for up to 8 hours per day, 11% received interrupted supplies for between 8 and 12 hours per day, and the supply to 6% was interrupted for more than 12 hours per day. During 1990–1995, 50% of the population received water with intermittence in distribution, compared with 35% of the population during 1985–1989.
Slovenia	Almost 120 000 people suffer interruptions in supply. Interruptions are most frequent in the summer and in rural areas. Organizational difficulties and financial problems are the most common causes.
Turkey	Discontinuous supply is reported in some areas.
Turkmenistan (Dashkovuz Region)	Water distribution is intermittent, with three periods of delivery (of 2 hours each) scheduled each day.

Sources: Mountain Unlimited *(93)*; Bertollini et al. *(94)*; Iacob *(95)*.

Box 4.2. Influence of network integrity and discontinuous supply on outbreaks of waterborne disease in Albania

Since 1988, Albania has had insufficient financial resources to maintain drinking-water quality and sanitation. There have been several known outbreaks of waterborne disease, and it is believed that water also contributes significantly to background rates of enteric diseases.

Although a number of aspects of water quality and supply need to be addressed, one of the principal problems is the poor state of Albania's drinking-water networks. High rates of water loss (up to 70%) occur, and this decreases the amount of drinking-water available and can compromise its quality by allowing ingress of contaminants. Although 100% of the population of Tirana, for example, has a networked supply, this figure is reduced to 73% in more rural areas. Public wells are used by 5%, public fountains by 22% and local springs by 0.7% of the population in these areas *(97)*.

Although Albania has high-quality groundwater sources available, the cost of pumping this water prevents a continuous supply. In the peripheral areas of Tirana, for example, 77% of the population has water supplied for only 2–3 hours per day and 33% of the population has water for 1 hour per day *(97)*. The poor condition of the supply network is further compounded by intermittent distribution at low pressure, which results in microbial contamination of the water in the supply system. High chlorine dosing is necessary to ensure that there is sufficient residual chlorine to secure a microbially safe supply. Nevertheless, 21% of the population receive insufficiently chlorinated water. This is attributed to financial restrictions and lack of specialized equipment. In addition, many of the raw water sources used are contaminated (coliform bacteria are found in 34% of water samples in Tirana), partly because treatment plants do not receive sufficient electricity *(98)*.

standards. Water supply is frequently interrupted, especially in regions of the southern Russian Federation, where water may only be supplied for a few hours per day. In Romania, supply is discontinuous in many regions and affects up to 60% of supply systems, leading to the distribution of water that is of dubious quality at best. Some interruptions exceed 12 hours per day. In Armenia, 50% of piped supplies do not meet quality standards. In a 5-year period more than 500 cases of dysentery and salmonellosis were reported from inadequate and poor-quality water supplies *(96)*.

Restrictions of supply can also be imposed in times of drought, as experienced in Romania and the United Kingdom, resulting in reduced access to water and, potentially, reduced quality caused by discontinuous supply.

AFFORDABILITY OF DRINKING-WATER

Charges made for supplied water reflect, to varying extents, the cost of abstracting, treating and supplying the water and maintaining the necessary installations, and the extent of subsidy by the state or municipality. Prices for supplying water (excluding any *pro rata* charge for treating wastewater) vary across the European Region, with those in western Europe varying from €53 per year in Rome to €286.6 per year in Brussels for a family in a house consuming 200 m³/year. Water charges in central European cities are lower, and vary from approximately €20 per year in Bucharest and Bratislava to €59 per

Fig. 4.4. Annual water charges for a household consuming 200 m³/year in selected European cities in relation to gross domestic product per person

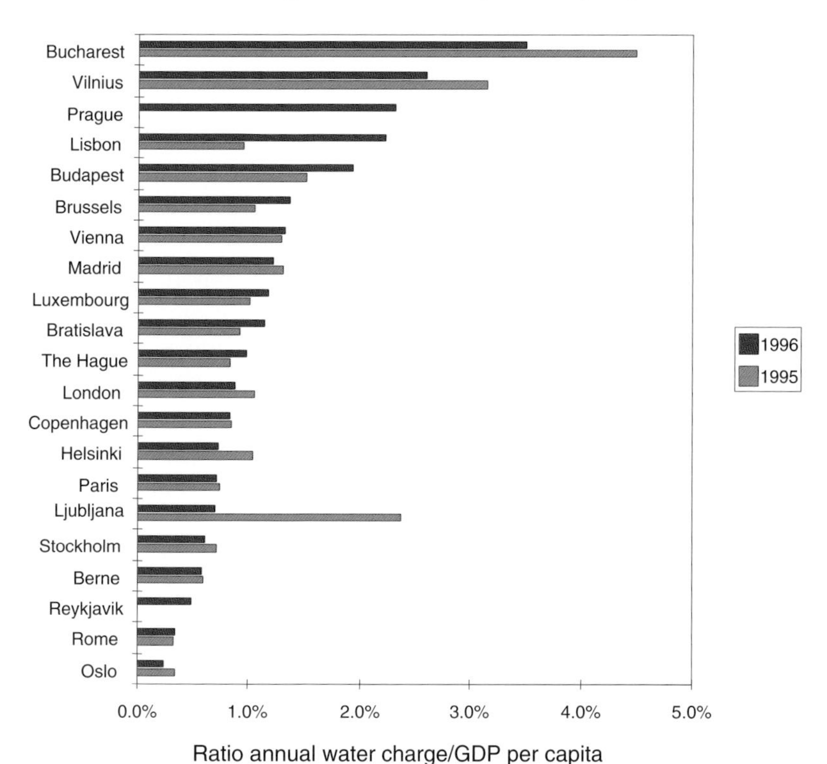

Source: International Water Supply Association *(99)*.

year in Prague *(99)*. However, when these annual charges are compared with the gross domestic product per person, the proportional costs in Bucharest, Vilnius and Prague are among the highest (Fig. 4.4).

Producing and providing clean water to consumers is expensive in initial capital outlay and the ongoing costs of maintenance, management and extension of services. Payment for water, however, is an emotive issue. Water is a basic human need, and the long-term sustainability of water supply requires that costs are recovered so that the infrastructure does not deteriorate, leading to a breakdown of the system. The costs of treating and disposing of wastewater should also be recovered.

In many countries in the eastern part of the Region, water prices have risen at a rate much higher than inflation since 1989, when state subsidies were reduced. Water storage and distribution systems in the eastern part of Germany and in Hungary, for example, are now fully financed from fees, with no state subsidy. The tariffs also include a portion for reconstruction and developing new schemes *(52)*. Tariffs remain low in some countries: Albania has a fixed charge of 5 lek (about €0.037) per m^3 for domestic consumers, and Turkmenistan has no charge *(93)*.

For all countries, cost containment should be an important objective of public utilities. Risk-taking and deficit-spending measures, based on high technology and the assumption that consumers or the government will pay at some future time, cannot be afforded and should be discouraged *(100)*.

Consumption of bottled water

Some countries in Europe have a tradition of consuming bottled water, especially mineral water (Table 4.3). Consumption of bottled water is increasing in these countries, and the market has widened to include other countries where bottled water is not traditionally consumed in large quantities. In Austria, France, Germany, Italy, Spain and Switzerland, consumption of bottled water increased by between 139% and 248% from 1983 to 1992 *(101)*.

Table 4.3. Consumption of bottled water in selected European countries in 1992 (litres per person)

Ireland	6
United Kingdom	9.5
Netherlands	15
Portugal	29
Spain	44
Austria	76
Switzerland	76
France	80
Germany	93
Belgium	105
Italy	116

Source: Bertollini et al. *(94)*.

The purchase of bottled water is largely a market phenomenon governed by societal customs. It may also indirectly indicate poor availability or quality (or perceived poor quality) of drinking-water from other sources *(89)*. However, the provision or purchase of bottled water is unlikely to be a cost-effective way of obtaining high-quality drinking-water. The cost per individual is estimated to be 2–5 times more expensive, money that could be better spent in ensuring a safe piped water supply *(102)*. The provision of bottled water may, however, be appropriate in certain situations. Emergency distribution in the case of serious contamination incidents affecting drinking-water supplies is an obvious example and, in some circumstances, the provision of bottled water to households with infants and young children may be the most appropriate action if supplies are contaminated with high levels of nitrate, in order to prevent methaemoglobinaemia (see Chapter 5).

Water quality

Sources and abstraction for drinking-water

Fresh water is abstracted from groundwater and surface water for a variety of purposes (see Fig. 3.2). The microbial quality of water is of primary importance, and the least polluted available source is generally preferred for potable supply. Groundwater is generally of better microbial and of more stable chemical quality than surface water, although at a local level some substances naturally occurring in groundwater, such as fluoride and arsenic, may be hazardous to human health. Shallow wells, in particular, are vulnerable to contamination. Upland surface waters are generally less contaminated than lowland ones.

The availability of natural resources largely determines the proportions of a country's drinking-water derived from surface water and groundwater. The convenience and practicality of using the sources (distance from the centres of population) are also influential. In some countries (Albania, Denmark and Turkmenistan, for example) almost all the drinking-water supply is provided by groundwater *(93)*, whereas in others the majority has to be abstracted from surface water. Throughout most of Latvia, over 50% of the drinking-water is derived from groundwater. About 50% of the rural population of Latvia use shaft or frame shallow wells (no deeper than 10–15 m), although in recent years the number of wells for collective use has decreased. In contrast, about 72% of the drinking-water in the United Kingdom and 87% of that in Norway is derived from surface water (Fig. 4.5). In Sweden, the largest cities use surface water, but nationally 50% of the population connected to a municipal area use surface water, 25% use surface water that has passed through gravel ridges and 25% use groundwater. Sweden also has about 400 000 private wells supplying permanent homes and between 200 000 and 400 000 wells supplying "recreation accommodation" *(103)*. In Estonia, 60% of drinking-water is derived from surface water or shallow wells and 40% from deep groundwater sources *(104)*.

Some countries have a policy of trying to reduce the proportion of drinking-water supplied from aquifers. This may be to redress the over-exploitation of aquifers that has occurred in the past, as in the Netherlands, or because of concern about contamination of groundwater, as in the Republic of Moldova *(24)*.

Drinking-water treatment

The type and degree of treatment required to make water wholesome differs according to the quality of the raw water source. Depending on the geology of the area, the groundwater may be

Fig. 4.5. Proportion of drinking-water derived from groundwater and surface water and desalination in selected European countries

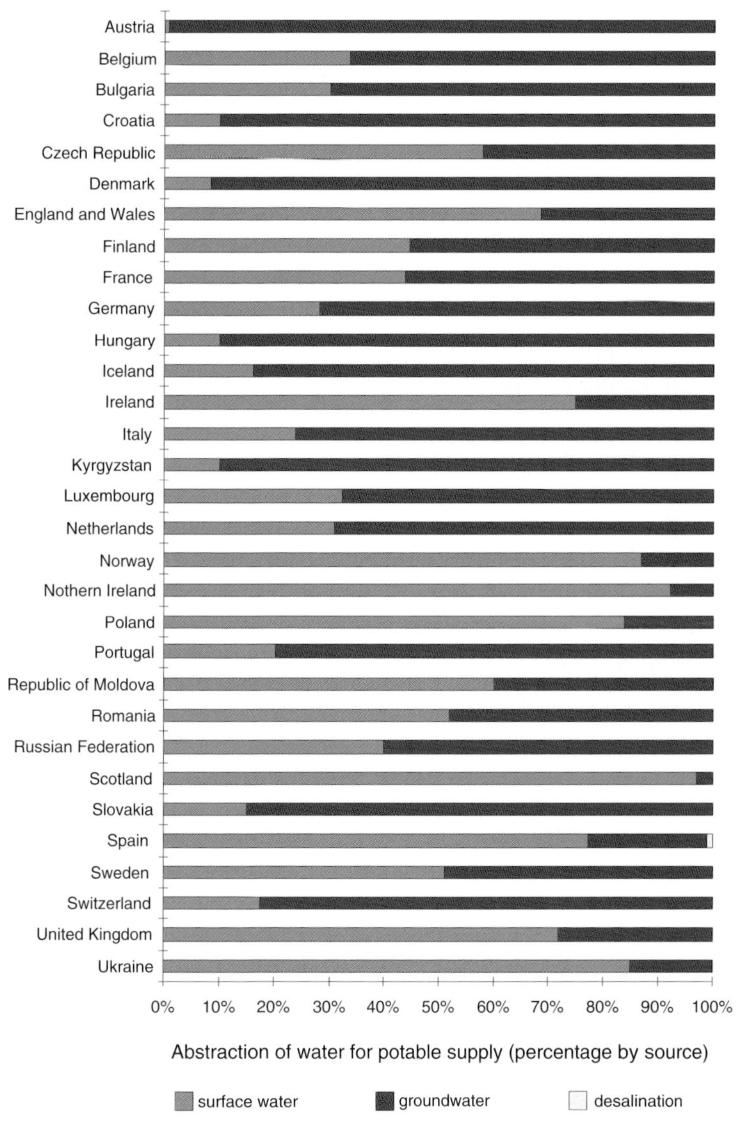

Note: In some countries, "groundwater" may include spring water from shallow sources. The estimated proportions of drinking-water derived from spring water are: Austria – 49%; Germany – 7%; Italy – 37%; Spain – 4%; and Switzerland – 46%. In addition, in Norway an estimated 6% of groundwater is bank-filtered surface water.

Sources: Eurostat *(64)*; Mountain Unlimited *(50)*; Water Research Centre, unpublished data (various).

contaminated with iron, manganese, carbon dioxide, fluoride or arsenic, and may require treatment and/or disinfection before use. Some groundwater has been affected by human activities and may be contaminated, for example, by nitrates, pesticides, solvents or pathogens.

The level of treatment required to ensure that poor-quality sources are suitable for consumption is significant and costly. Consequently, private water supplies, especially those supplying a single dwelling, may receive very limited or no treatment. Preventing pollution rather than intervening technologically to remove it is generally preferable. However, in many cases a legacy of historical pollution has to be addressed, especially as competing demands on water resources may result in poorer-quality sources being used for drinking-water.

Satisfactory treatment of water for potable supply and maintenance of the distribution networks are compromised in many European countries by financial limitations or a shortage of human or technical resources. Many countries in the eastern part of the Region report such problems, along with organizational difficulties. Unusually among the countries in western and northern Europe, authorities in Sweden report a personnel problem, in that human resources at waterworks have been reduced to a level of concern (Table 4.4). Financial constraints on the water supplies for small communities appear to be common.

Many countries in Europe use bank filtration as an economical method of improving the quality of surface water before abstraction for drinking-water. This reduces the need for conventional treatment of the water by removing substantial particles and microbial and chemical contamination. It is not practical in all cases, however, as it depends on suitable geology (Box 4.3).

One of the emerging concerns about the treatment of water is the production of disinfection by-products that have been shown to be carcinogenic in animal studies (Box 4.4). Most disinfectants are oxidizing agents and can react with natural compounds dissolved in the water to give products that are potentially of concern to human health. These by-products are more likely to be formed in surface waters, as these contain higher concentrations of organic matter such

Table 4.4. Countries in the European Region reporting financial, human resource or organizational problems compromising the quality of drinking-water in 1997

Country	Main difficulties experienced compromising water quality
Albania	Financial constraints. Organizational, technical and human resource problems reported. Old supply systems suffering corrosion and old, manual chlorination equipment are highlighted
Belgium (Wallonia)	Financial constraints for water treatment (cost per m^3) in areas where small numbers of people are connected to many small sources
Croatia	Financial constraints
Czech Republic	Financial constraints prevent the use of the best available technology
Estonia	Financial constraints
France	Financial constraints, especially in small communities of less than 100 inhabitants
Greece	Financial constraints
Lithuania	Financial constraints. Unavailability of equipment and chemicals
Malta	Financial constraints
Republic of Moldova	Financial constraints and poor availability of equipment
Romania	Training of personnel. Financial considerations restrict improvement in equipment for water treatment and chlorination and for correcting the overloading of water treatment capacities
Slovenia	Organizational difficulties and lack of human resources. Financial constraints have resulted in a lack of sophisticated treatment plants
Sweden	Human resources in waterworks have been reduced

as humic and fulvic acids. However, the risks appear to be small, and the overriding priority in providing clean drinking-water must be the microbial quality, in order to prevent waterborne infectious diseases *(105)*. It is therefore important to establish a balance

Box 4.3. Bank filtration

Bank filtration is extensively used in a number of European countries, including Germany, Hungary and the Netherlands. Water from a surface water source, usually a river, is allowed to filter into the groundwater zone through the riverbank and to travel through the aquifer to an extraction well some distance from the river. In some cases there is a short residence time in the aquifer, perhaps as little as 20–30 days, and there is almost no dilution by natural groundwater *(44)*.

Bank filtration is often very effective in improving the microbial quality of water, and also removes gross contamination such as suspended solids. Some chemical contaminants may also be removed by adsorption and complex formation in the filtration medium, although the extent of removal depends on the chemical involved and the geology of the area. Conversely, certain contaminants may be mobilized, resulting in an increased concentration. This is especially so in agricultural areas, where the concentrations of pesticides and fertilizers may be high. Depending on the geology, the quantities of elements such as iron and manganese may also be increased, particularly under anoxic conditions.

Bank-filtered water is often classified as groundwater, despite its recent origin in a surface source. This can produce problems in interpreting data on the relative quality of groundwater and surface water, since many of the assumptions regarding groundwater (such as low levels of contaminants and a long recharge time) may not apply to some bank-filtered water, depending on the quality of the surface water, the geology of the area and the residence time before abstraction.

between maintaining effective disinfection and the need to reduce disinfection by-products to an acceptable level, which clearly favours chemical disinfection.

The prevailing treatment and disinfection methods in European countries are influenced by the quality of source water, financial resources, available technology and historical practice. In the newly independent states, chlorine disinfection is most commonly used, often by an indirect chlorine gas procedure (Box 4.5) *(107)*. Ultraviolet radiation is commonly used in some countries of Europe, for example with groundwater from the alpine regions of Liechtenstein (where faecal contamination from cattle is a concern) and for the 6% of the Icelandic population whose supplies are derived from surface water.

A number of different approaches are normally used within the same country (Table 4.5). In Poland, chlorination is the main method of

Box 4.4. Disinfection procedures and by-products

Chlorine is the most widely used drinking-water disinfectant in Europe. The reaction of chlorine with water results in hypochlorous acid, which dissociates to produce the hypochlorite ion. Alternatively, sodium hypochlorite can be added as a disinfectant. Chlorine reacts with natural organic chemicals dissolved in the water (humic and fulvic acids) to form chloroform (trichloromethane). In addition, hypochlorous acid oxidizes bromide present in the water to form hypobromous acid, which also reacts with humic and fulvic compounds to produce bromoform. Mixed trihalomethanes (dibromochloromethane and bromodichloromethane) are also produced.

Animal studies using high doses of these compounds have indicated their carcinogenic potential, and many countries regulate their levels in drinking-water. There are other disinfection by-products, some of unknown toxicity, and the presence of trihalomethanes is considered to provide a broad indicator for these other chlorination by-products. Other by-products include chlorinated acetic acids, chloral hydrate, chloroacetones, halogenated acetonitriles, cyanogen chloride and chloropicrin *(105)*. Some other disinfection by-products, especially chlorophenols, impart an unpleasant taste and odour to the water.

Ozone is often used as an alternative where the natural water contains substances that would produce adverse taste and odour if chlorine were used. This is a preferred method of disinfection in some countries (such as France and Germany) and has been shown to be effective against *Giardia* cysts and *Cryptosporidium*. It is expensive, however, and there is no residual disinfection action within the water supply mains. Biological growth may subsequently occur within the piping. To prevent this, low-level chlorination is often used after ozonation to impart a residual disinfectant action. A by-product of ozonation is bromate, formed by the oxidation of the bromide present in water, which has been shown to be carcinogenic in laboratory animals.

Chlorine dioxide is a further option, used in Belgium, France, Germany and Italy. Although fewer trihalomethanes are produced with chlorine dioxide than with chlorine, the concentrations of chlorite and chlorate may be higher in some water. Sodium chlorite has no known adverse physiological effects on humans, but methaemoglobinaemia, anuria, abdominal pain and renal failure are associated with chlorate poisoning. Chlorine dioxide is more effective than chlorine in inactivating *Giardia* cysts and *Cryptosporidium* oocysts, but the residual is unstable *(106)*.

Ultraviolet radiation is used to disinfect water supplies, alone or in conjunction with chemical disinfectants. As with ozone, the disadvantage is the lack of residual disinfection in the distribution, and chlorination may therefore also be required. It is effective against bacteria and many viruses, but not against *Giardia* cysts and *Cryptosporidium* oocysts *(106)*.

water disinfection in public water supply systems. In wells and individual water supply systems, chloramine and chlorinated lime (containing up to 30% of active chlorine) are commonly used. Chlorine dioxide is in wide use in France, Germany and Italy, but ozone is

Box 4.5. Drinking-water problems in the newly independent states

For various reasons, including the pollution of surface water and sometimes groundwater, providing drinking-water in sufficient quantities and of sufficient quality has become an acute problem in many regions of the newly independent states. The countries with the largest percentages of water samples that fail to comply with national standards are Azerbaijan, Estonia, Kazakhstan, Kyrgyzstan and Turkmenistan. Drinking-water has historically been abstracted mainly from surface resources. The most severely affected areas in terms of tap water quality are in regions of central Asia around the Aral Sea. Up to 60% of the water samples from municipal water supply systems and approximately 38% from rural supply systems in the Kyzyl-Orda region do not comply with national standards. Established water quality standards are not always met because of the high degree of pollution, inadequate treatment, and secondary bacterial pollution in the supply networks. About one third of the population uses water from untreated local sources. These factors have contributed to an increase in outbreaks of acute digestive infections. In 1993 there were 17 outbreaks, including 10 of dysentery. An outbreak of gastrointestinal infections and typhoid was also recorded in 1993 in Rostovskaya *oblast*, where 300 people became ill in Volgodonsk.

Sources: State Committee of the Russian Federation on Environmental Protection *(9);* Abakumov & Talayeva *(23).*

preferred at some sites in France and Germany and is also common in the Netherlands.

Water quality may be threatened (through recontamination) where discontinuous chlorination occurs – Albania, Armenia, Estonia, Greece, Lithuania, the Republic of Moldova, Romania, Turkmenistan and Ukraine reported such problems *(50,93)*. In many cases this is attributed to old, broken or manual equipment or a lack of chlorine *(108)*. Automatic equipment for continuous chlorination using sodium hypochlorite has now been installed in a number of cities in Albania.

Insufficient treatment of water, especially disinfection, is a particular problem in small supplies. Many small waterworks in Norway, for example, do not have sufficient disinfection, and problems of microbial contamination are reported in small supplies in the Wallonia region of Belgium because of discontinuous chlorination or no disinfection. Achieving high-quality drinking-water (microbial and chemical) in small communities of less than 100 inhabitants, at a reasonable price, is also regarded as a problem in France.

Table 4.5. Main types of treatment used for water for potable supply

Country	Groundwater and spring water	Surface water
Belgium	Aeration and disinfection only Few granulated active carbon and nitrate removal units Some iron and manganese removal Some air-stripping for organics	Chemical coagulation (flocculation, rapid sand filtration, O_3 + granulated active carbon filtration) and disinfection
Finland	Mostly alkalization. Iron and manganese removal and some disinfection. Some O_3 + granulated active carbon treatment	Chemical coagulation, clarification, filtration and disinfection. Some O_3 + granulated active carbon treatment
France	Disinfection only Some nitrate removal (ion exchange and biological denitrification)	Chemical coagulation, O_3+ granulated active carbon (also some advanced oxidation processes, such as O_3+ granulated active carbon + O_3/H_2O_2) and disinfection Few waterworks with membrane technology Some nitrate removal (mainly ion exchange)
Germany	Most groundwater not treated (some disinfection) except where pesticide, solvent or nitrate removal is required	Bankside filtration commonly used Activated carbon in common use Recharged groundwater/bank filtration combined with coagulation, filtration, O_3 + granulated active carbon, disinfection
Iceland	No disinfection	Filtration and ultraviolet radiation No supplies are chlorinated
Italy	Little or no treatment, mainly disinfection Considerable use of granulated active carbon for pesticide, organic solvent removal Trend away from chlorine to chlorine dioxide for disinfection	Traditional physical or physical and chemical treatment Also complex treatment such as granulated active carbon and disinfection Increasing use of chlorine dioxide
Liechtenstein	Majority enters supply without treatment or disinfection In Alpine regions, filtration or ultraviolet radiation is employed	
Netherlands	Aeration and multistage sand filtration	Extensive use of multistage treatment, including dune infiltration, coagulation, activated carbon and disinfection with chlorine or ozone (trend away from chlorine use because of trihalomethane formation) Concern to maintain low assimilable organic carbon in distribution

Table 4.5. (contd)

Country	Groundwater and spring water	Surface water
Republic of Moldova	Groundwater is used for drinking-water without treatment in about 50 towns	
Russian Federation	Disinfection (chlorination) only	
Slovakia	75% supplied following disinfection only (especially in southwestern and central Slovakia) Remainder treated for removal of Fe, Mn, NH_4, CO_2, oxidizability and methane	Six largest plants use chemical coagulation and chlorination or chloramination Stream abstractions often use sand filtration or slow sand filtration
Spain	Minimal treatment, mainly disinfection only Filtration/chemical coagulation and granulated active carbon or O_3 + granulated active carbon sometimes used	Most commonly chemical coagulation with rapid filtration Granulated active carbon or O_3 + granulated active carbon also relatively frequently used Chlorine widely used and high doses are often required, leading to concern over trihalomethanes
Sweden	None	Treatment and disinfection
Turkmenistan	Disinfection by chlorination only	Disinfection by chlorination only
United Kingdom	Disinfection only, using chlorine Iron and manganese removal for some sources Approximately 20 waterworks with nitrate removal (all ion exchange) Removal of organics (pesticides and solvents) by O3 + granulated active carbon on 20% of supplies	Mostly chemical coagulation and disinfection Some slow sand filtration Removal of pesticides by granulated active carbon or O_3 + granulated active carbon for one third of supplies

Sources: Mountain Unlimited *(50,93)*; WHO Regional Office for Europe *(102)*.

Drinking-water distribution

Effects of distribution on water quality

Microbial contamination

Microorganisms found in drinking-water distribution systems may enter the water through faulty source protection, treatment or disinfection, or by recontamination of pipes through back siphoning or regrowth. When the water pressure in the mains is not sufficient, such as when water supplies are discontinuous, microorganisms can enter the distribution system through leaks and contaminate the water. This is most serious when mains pipes are laid alongside sewerage systems. Loss of pressure may also result in back-siphoning of water through the plumbing system from sources of contamination such as taps. In Albania, the corrosion of old pipework also contributes to the contamination of drinking-water.

The growth of bacteria in distribution systems can result in discoloration and the imparting of taste and odour to the water. It can also lead to the proliferation of microorganisms and even higher organisms such as *Asellus aquaticus* (water louse). The growth of most bacteria is limited by the concentration of assimilable organic carbon in the water and/or the concentration and type of disinfectant residual. Even where disinfection is practised, biofilm slimes can harbour bacterial and protozoal pathogens that can be released into the distribution system when parts of the biofilm slough off. This can occur during and after chlorination. Pathogens may also be protected from the action of chlorine (and other disinfectants) in the biofilm. The presence of biofilms may also initiate and promote corrosion in water distribution systems. This may be considered an indirect risk if the corrosion leads to a failure of the pipe system and ingress of material from the surrounding environment.

Conventional water treatment processes such as coagulation, sedimentation and filtration can reduce assimilable organic carbon, but concentrations are increased by oxidants such as chlorine and ozone. To achieve the very low concentrations required to limit regrowth in the absence of a residual disinfectant, multi-stage treatment is required, including some form of biologically active process such as slow-sand or granulated active carbon filtration with intermediate stages of oxidation *(109)*.

Despite efforts to minimize carbon sources, concentrations can still be sufficient to allow the regrowth of microbes in long distribution systems. Finland, for example, reports such problems.

Chemical contamination

The materials used in constructing distribution systems, the integrity of the systems and the maintenance of positive water pressure within the system can all affect the quality of the water supplied. The choice of pipes and materials depends on the size of the pipes and has changed with time. Large trunk mains have historically been constructed of iron, steel or asbestos cement. Asbestos cement has now largely been phased out and plastics such as unplasticized polyvinyl chloride and polyethylenes such as medium-density polyethylene are supplementing the use of iron and steel. Nevertheless, there are concerns about the leaching of antioxidants, stabilizers and plasticizers from these products into water and leaching tests may be carried out on materials to be used in contact with drinking-water. Contamination can be better controlled by focusing on the materials rather than on the water quality.

Until the middle of the 20th century, lead was favoured as the construction material for service pipes in many countries. However, because of the recognized need to reduce exposure to lead and, in particular, concerns about the effects of dissolved lead on the development of the nervous system in children, lead piping is no longer installed in many countries and is gradually being removed from the distribution system. Newly laid polyvinyl chloride pipes with lead as a stabilizer are reported to contaminate drinking-water supplies with lead, although such leaching appears to be a short-term problem. The WHO guideline value *(105)* and the new EU standard *(48)* for lead in drinking-water are both 10 µg/l; the previous EU standard was 50 µg/l. The main remedy is to remove lead piping, which is expensive and time-consuming, and it may therefore take time before drinking-water in general meets the standard. The EU allows until 2013 for implementation to be completed.

In the past, iron distribution pipes were often coated internally with coal tar to reduce corrosion. Such linings contain relatively high levels of polycyclic aromatic hydrocarbons, a number of which are carcinogenic. The smaller, more water-soluble polycyclic aromatic hydrocarbons such as fluoranthene leach into the carried water. Although these polycyclic aromatic hydrocarbons are generally not regarded as carcinogenic, the use of such linings is now widely prohibited. Metal pipes now often receive instead an internal coating of bitumen, which contains very low levels of polycyclic aromatic hydrocarbons.

Leakage

The efficiency of water transfer within supply networks directly affects water demand. Leakage in water distribution networks may be significant and can account for over 50% of the water entering the network (Table 4.6). However, poor metering and monitoring in some countries makes accurate estimation difficult.

Leakage may be reduced by a number of methods such as:

- repairing visible leaks;
- establishing leakage control zones;
- becoming aware of, locating and repairing leaks not visible from the surface;
- conducting telemetry of zone flows;
- reducing pressure;
- replacing mains;
- subsidizing the detection and repair of leaks in supply pipes for domestic customers or businesses;
- repairing leakage through the maintenance of service reservoirs;
- minimizing service reservoir overflow losses; and
- detecting and repairing leakage in trunk mains.

Estimates of future network efficiency in France range from 78% in urban areas and 72% in rural areas to 80% for both. Nevertheless, improvement is not the overall trend throughout Europe.

Many countries are attempting to reduce leakage rates, often encouraged by government concern. Water companies in the United Kingdom have statutory targets (with financial penalties) for reducing leakage, and this has proved successful in some regions (Table 4.7). Yorkshire Water Services in the United Kingdom considers that there is little scope for further reductions by repairing visible leaks and establishing leakage control zones. Nevertheless, options such as telemetry of zone flows, reducing pressure, subsidizing the detection and repair of customer supply pipes, repairing leakage through the maintenance of service reservoirs and replacing mains could bring about further reduction *(52)*.

A potential conflict of interest has arisen recently in the United Kingdom, as reduction in mains pressure to reduce leakage has raised

Table 4.6. Estimated losses from water networks in selected European countries, mid-1990s

Country	Estimated losses from water networks
Albania	Up to 75%
Armenia	50–55%
Bulgaria	
Sofia	30–40%
Other than Sofia	More than 60%
Croatia	30–60%
Czech Republic	33%
France	30%
National average, 1990	
Paris	15%
Highly rural area	32%
Germany (western Germany)	3700 litres per km of mains pipe per day 112 litres per property per day
Hungary	30–40%
Italy	
National average	15%
Rome	31%
Bari	30%
Kyrgyzstan	20–35%
Republic of Moldova	40–60%
Romania	21–40%
Slovakia	27%
Spain	
Settlements > 20 000 population	20%
Madrid	23%
Bilbao	40%
Ukraine	About 50%
United Kingdom (England and Wales)	8400 litres per km of mains pipe per day 243 litres per property per day

Sources: Mountain Unlimited *(50,93)*; Water Research Centre *(110)*; Istituto di Ricerca sulle Acque *(111)*.

Table 4.7. Reduction in leakage (Ml/day) by Yorkshire Water Services in the United Kingdom from 1994–1995 to 1996–1997

	1994–1995	1996–1997
Total leakage	536	420
Supply pipe losses	101	98
Distribution losses	435	322

Source: Krinner *(52)*.

concerns from fire-fighting services that the pressure in some areas may now be insufficient to fight large fires effectively.

Reuse and recycling

The practice of reusing wastewater is increasing in EU countries, primarily to alleviate the lack of water resources in certain regions such as southern Europe. This is addressed in article 12 of the Directive on urban wastewater treatment *(60)*, which specifies that treated wastewater shall be reused whenever appropriate.

Rural reuse (irrigation)

The largest application of direct wastewater reuse in Europe is for irrigation of crops, golf courses and sports fields, and concerns have been expressed that pathogens from the wastewater may come in contact with the public. Epidemiological studies have shown that crop irrigation with wastewater causes a significant increase in intestinal nematode infections in crop consumers and field workers when the wastewater is untreated, but not when it is adequately treated before use *(56)*. Wastewater irrigation is successfully practised in parts of France, Germany, Portugal, Spain *(56)* and Poland. In Portugal, the bacterial quality of lettuce irrigated with wastewater was found to be three orders of magnitude better than lettuce irrigated with river water *(112)*. Other countries in Europe that do not practise wastewater irrigation commonly import produce and flowers that have been irrigated with wastewater.

Most European countries do not have specific regulations on wastewater reuse, although regulations and guidelines regarding the uses of water may apply. WHO *(57)* has developed quality guidelines for the use of treated wastewater in irrigation. These specify the following.

- Treated wastewater to be used for restricted irrigation (that is, the irrigation of all crops except those eaten uncooked), should contain no more than 1 human intestinal nematode (human roundworm, whipworm or hookworm) egg per litre.

- Treated wastewater to be used for unrestricted irrigation (that is, the irrigation of crops eaten uncooked) should contain no more than 1 human intestinal nematode (human roundworm, whipworm or hookworm) egg per litre and should contain no more than 1000 faecal coliform bacteria per 100 ml.

Effluents complying with standards can be produced by treating wastewater in waste stabilization ponds *(113)*. Conventional treatment processes, such as activated sludge, can achieve the guideline on nematodes because of the two periods of primary and secondary sedimentation, but a tertiary treatment process such as maturation ("polishing") ponds, ultraviolet light or chemical disinfection is required to meet the guideline on faecal coliform bacteria.

Spray irrigation using treated wastewater is not recommended, as it may constitute a risk to operators and adjacent communities by the inhalation of pathogens in aerosol droplets.

The potential for water reuse and recycling has not been fully exploited in many areas, and economic or regulatory incentives are likely to be required to encourage its use. Experience in water reuse is therefore lacking in much of Europe. In many circumstances, the limiting factors can be the quality of the water available, the potential hazards for secondary users and public perception.

Desalination

The desalination of seawater costs approximately €0.7 per m^3, including energy cost and depreciation, and its contribution to the total water supply in Europe is very limited. In Monaco and Spain, desalination contributes 0.45% and 0.33% of the water supply, respectively. In Malta and the Balearic Islands, characterized by a comparatively dry climate and relatively limited surface water resources, desalination is of greater importance, contributing 46% of the total water in Malta *(114)* and one third of the total urban water supply (91 500 m^3/day) in the Balearics *(52)*.

The viability of desalination as a more widespread option for the future will depend on technological advances, the cost of energy and the cost of using alternative sources.

WATER TRANSFER

There are several large-scale, inter-basin water transfer schemes in Europe. The Rhône-Languedoc transfer and the Canal de Provence in France, with capacities of 75 and 40 m^3/s, respectively, are two of the largest. Smaller schemes exist in Belgium, Greece and the United Kingdom. Spain has about 50 small inter-basin water schemes able to transfer about 1.5 km^3/year. For inter-basin transfers to be an efficient, cost-effective and acceptable means of satisfying water demand in regions with low water resources, the environmental sustainability and economic viability need to be assessed carefully *(52)*. Schemes to carry water from one catchment area to another have encountered considerable resistance from local populations in some regions, especially if there is evidence or suspicion of water shortage in the donor region.

The removal of large volumes of water from natural watercourses for water transfer may have detrimental environmental effects and affect water availability in downstream regions. Such effects were experienced in Turkmenistan and Uzbekistan, where water has been diverted from the two major rivers that feed the freshwater Aral Sea to areas that were being developed to allow intensive cotton production. These water transfers have seriously harmed water resources within the basin. The sea has shrunk significantly, and the quality of the rivers and groundwater in the area has seriously deteriorated (see Box 3.1).

5

Health effects

Restricted access to drinking-water

Access to a sufficient supply of safe water is essential in maintaining public health. Situations with inadequate water directly and indirectly affect health. Poor hygiene caused by the lack of water results in the increased transmission of infectious diseases *(115)*. Where the sources of potable water are of poor quality, or the financing, staffing and other infrastructure to maintain the distribution system are lacking, mortality rates attributable to infectious diseases, to the availability of sanitation services and to general hygiene may increase. Inadequate water supplies increase the likelihood of person-to-person disease transmission and can compromise the effectiveness and efficiency of water-based sewage collection and treatment processes, posing an additional risk of disease.

In recent years the privatization of water supplies in some countries has resulted in an increase in the number of households disconnected from water supplies. In the United Kingdom, for example, the number of domestic disconnections rose from 8000 to over 21 000 between 1989 and 1992 following privatization of the water industry. Water has become an increasingly expensive commodity and evidence indicates that, where water meters have been introduced, those with lower incomes use less water *(116)*. Few studies have examined the social and health effects of water disconnection or of excessively low water use to save money *(117)*. One study showed a significant correlation between the number of disconnected households and the incidence of hepatitis A and shigellosis in some areas of the United Kingdom. This study also implicated involuntary reduction in water use because of economic deprivation as the cause of the increase in disease *(118)*.

DRINKING-WATER QUALITY

The quality of drinking-water in supply depends on many factors, including the quality of the raw water source, the extent and type of treatment and disinfection used, the materials and integrity of the distribution system, and the maintenance of positive pressure within the network. Excessive abstraction and mineral excavation can also lead to contamination of the groundwater supply. In Kyrgyzstan, the groundwater supply has suffered saline intrusion and toxin contamination. Surface water is of equally or even poorer quality in Kyrgyzstan. Spring floods lead to annual outbreaks of hepatitis, typhoid and diarrhoeal disease *(96)*.

A number of chemical compounds are produced during disinfection of drinking-water, some of which have been shown to be carcinogens in animal studies. Standards and/or guidelines have been set for known by-products, but the toxicity of by-products that have not yet been identified is also of potential concern. The microbial safety of drinking-water is the overriding concern, and should not be compromised because of concern about possible risks to health from the presence of disinfection by-products.

Microbial quality

The provision of a microbially safe drinking-water supply is the most important step that can be taken to improve the health of a community, by preventing the spread of waterborne disease *(105)*. Monitoring the microbial quality of drinking-water is therefore aimed at verifying that it is free from such contamination. The pathogenic organisms directly responsible for the spread of disease are of concern, but detecting them is difficult, expensive and time-consuming. Pathogen detection is therefore not appropriate for routine monitoring. In addition, not all pathogens have as yet been characterized and methods for detection of some others remain unavailable.

Instead, water is examined for bacteria that indicate the presence of faecal contamination *(105)* (Box 5.1).

Box 5.1. Microbial indicators of faecal contamination

The ideal indicators of faecal contamination should be universally present in the faeces of humans and warm-blooded animals and should not grow in natural bodies of water. They should be easy to detect and enumerate in water. Their persistence and removal during water treatment should be similar to those of waterborne pathogens, so that they not only act as indicators of faecal contamination but also monitor the effectiveness of any water treatment measures in removing pathogens from the supply *(105)*.

Escherichia coli and the thermotolerant ("faecal") coliform bacteria are the organisms most commonly used to indicate faecal contamination. Faecal streptococci are more persistent than *E. coli* and coliform bacteria; they are monitored regularly in fewer European countries than coliform bacteria and may be used as a supplementary measure to investigate supplies where coliform bacteria measurements have indicated microbial contamination. Faecal streptococci have been shown to be good indicators of microbial contamination of a network system with discontinuities in water supply *(119)*.

WHO guidelines for drinking-water quality recommend that indicators of faecal contamination (*E. coli* or thermotolerant coliform bacteria) should not be detectable in any 100-ml sample of any water intended for drinking *(105)*. For treated water entering the distribution system, neither faecal indicators nor total coliform bacteria should be detectable in any 100-ml sample. For water within the distribution system, the recommendation is again that no faecal indicators should be detectable in any 100-ml sample. The same applies to total coliform bacteria, although the guideline indicates that, for large supplies for which a large number of samples are examined, total coliform bacteria should not be present in 95% of samples taken throughout any 12-month period *(105)*. Fig. 5.1 and 5.2 show the percentage of drinking-water samples exceeding the national standards for total and faecal coliform bacteria (the national standards are specified in the figures).

Many European countries have standards for both total coliform bacteria and faecal indicators of zero per 100 ml. Although no countries permit the supply of drinking-water containing faecal indicators, a number of countries in the eastern part of the Region permit some evidence of microbial contamination, as detected by total coliform bacteria (0–3 per 100 ml).

Fig. 5.1. Percentage of drinking-water samples exceeding national standards for total coliform bacteria in 1995[a]

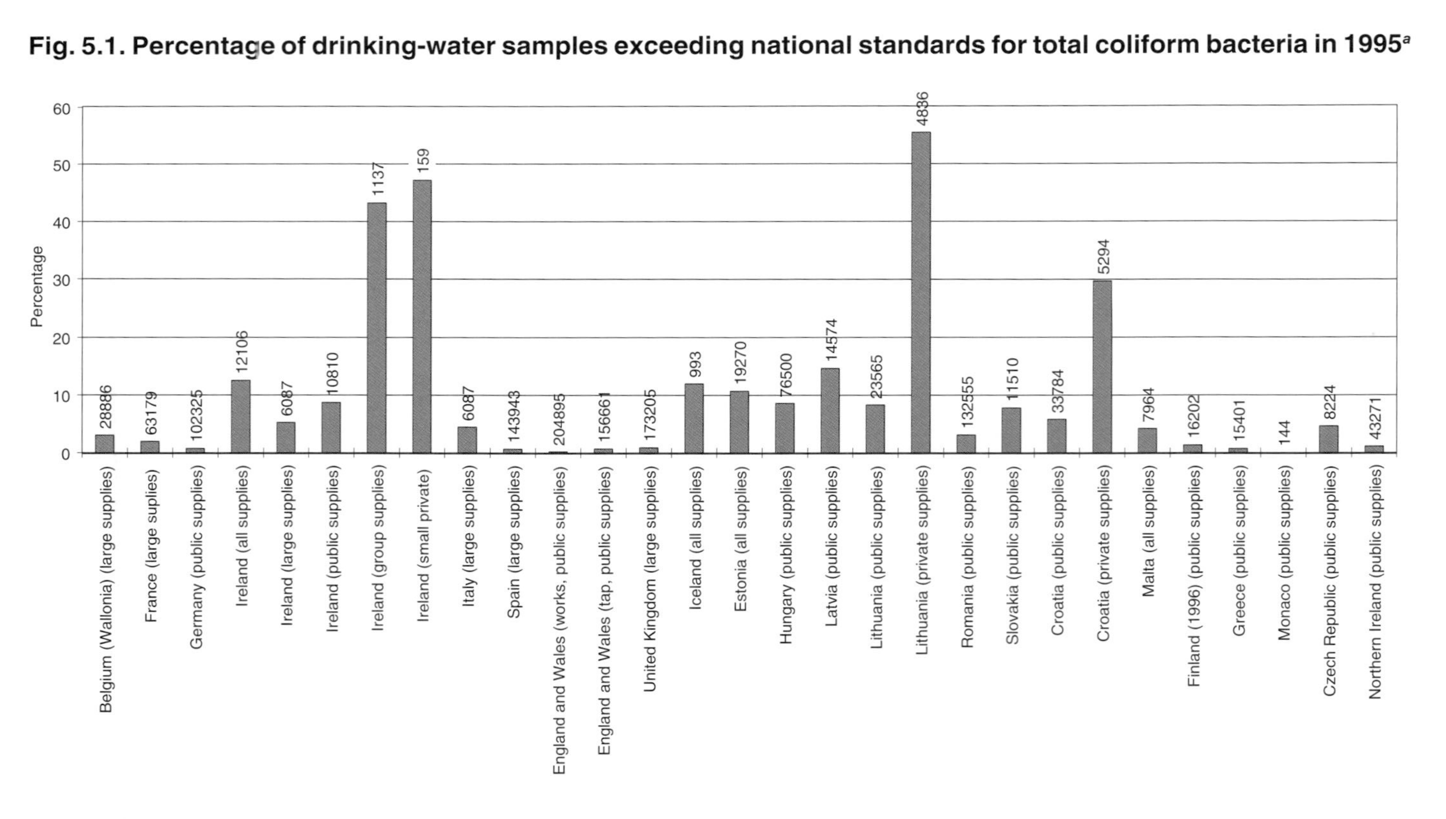

[a] Values above each bar are the numbers of samples analysed for total coliforms.

Fig. 5.2. Percentage of drinking-water samples exceeding national standards for faecal coliform bacteria in 1995[a]

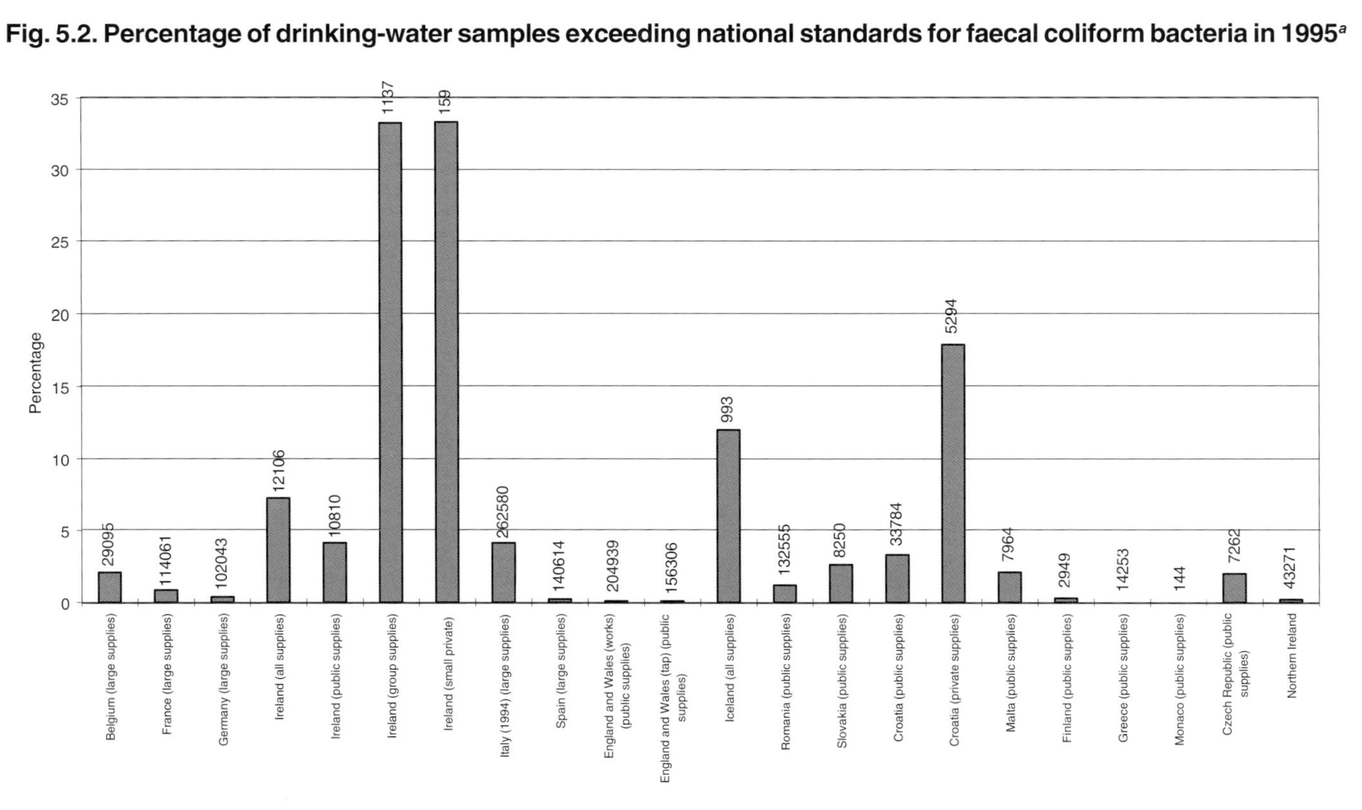

[a] Values above each bar are the numbers of samples analysed for faecal coliforms.

Community-managed supplies may not be subject to such stringent standards as public supplies, and the quality of community-managed supplies may therefore be compromised. In Lithuania, for example, the standard for total coliform bacteria in public supplies is 0.3 per 100 ml, whereas 1 per 100 ml is permitted in community-managed supplies. Microbial contamination is reported to be a particular problem in private wells. Community-managed supplies may not be examined as part of a routine monitoring programme and may only, as in Ireland, be examined on an emergency basis in response to reported problems. Thus samples will not be representative of the supplies as a whole, and the reported proportion that exceed the standard will be higher than would be the case for random sampling.

Indicators of microbial contamination are among the parameters found most frequently at levels of concern in the drinking-water of many European countries. These include Andorra, Croatia, the Czech Republic, Estonia, Germany, Greece, Iceland, Liechtenstein, Malta, Norway, Slovakia, Slovenia, Sweden, Turkey, Turkmenistan and Ukraine. In parts of Belgium and France, microbial contamination is reported to be a particular concern in small supplies.

The bacterial quality of drinking-water depends on a number of factors – the depth of the aquifer, the condition of the distribution network, the efficiency of treatment and disinfection in particular. Diseases can also be contracted by exposure to microbially contaminated water during recreation, and potentially as a result of the use of contaminated water for irrigation. Spray irrigation, in particular, is a potential hazard to agricultural workers, and disease can be spread via foodstuffs, especially produce eaten uncooked, that have been irrigated with water containing pathogens.

Agricultural activities, often coupled with poor source protection, are widely regarded as contributing to microbial contamination of water sources; this is reported as a factor in Liechtenstein. Where the water source is open to animals or there is widespread use of septic tanks, the drinking-water supplies are more likely to be contaminated *(120)*.

Poor sewerage systems and the discharge of untreated sewage are likely to affect source water quality *(92)*. Some organisms, such as

Cryptosporidium oocytes *(121)*, are more resistant to treatment than others and can be discharged into surface waters in sewage effluent. Lack of chemical pretreatment, improper backwashing procedures, poor application of raw water to the filter and failure to monitor plant conditions have been implicated in outbreaks of waterborne giardiasis *(122)* and cryptosporidiosis *(123,124)*.

Public supplies may be at risk if financial or technical constraints result in discontinuous chlorination. Technical faults and faulty connections may result in wastewater infiltrating into the supply network, potentially contaminating the drinking-water. Individual wells are vulnerable to contamination from adjacent sewage ditches. This is considered to have caused between 2 and 10 outbreaks of disease in small communities each year in Sweden *(103)*.

Infections

A number of serious infectious diseases, such as hepatitis A, cholera and typhoid fever, can be spread via contaminated drinking-water, as can more common intestinal diseases such as gastroenteritis. Bacteria cause cholera (*Vibrio cholerae*), typhoid fever (*Salmonella typhi*), bacillary dysentery or shigellosis (*Shigella* spp.) and campylobacteriosis (*Campylobacter* spp.). Viruses cause others, such as hepatitis. In addition, protozoa cause parasitic diseases.

Waterborne infectious diseases not only cause preventable illness and death but may also have substantial economic effects on the affected people and their families and society as a whole, including expenses for health care and loss of productivity.

Surveillance

Available data on waterborne diseases and outbreaks are often incomplete and inconsistent. Most diseases that can be spread by water are also spread through faecal contamination by other routes such as person-to-person contact and on contaminated food. Recorded cases of diseases could therefore have resulted from any of these routes of infection. Differences in recording (Table 5.1) and reporting procedures, disease classification and financial restrictions, and variation in the legal basis for reporting between countries, also

Table 5.1. Countries in the European Region indicating whether they keep records of seven waterborne diseases

Country	Gastro-enteritis	Ameobic dysentery	Bacillary dysentery	Cholera	Crypto-sporidi-osis	Giardiasis	Typhoid fever
Albania	Yes			Yes			Yes
Andorra	Yes	Yes	Yes	Yes	No	No	Yes
Austria	No	Yes	Yes	Yes	No	No	Yes
Belgium							
Flanders	Yes[a]	No		Yes	Yes	Yes	Yes
Wallonia	No	No		Yes	Yes	Yes	Yes
Croatia	Yes	No	Yes	Yes	No	No	Yes
Czech Republic	Yes	No	Yes	Yes	Yes	Yes	Yes
England and Wales	Yes	Yes	Yes	Yes	Yes	Yes	Yes
Estonia	Yes	Yes	Yes	Yes	Yes	Yes	Yes
Finland	No	Yes	Yes	Yes	Yes	Yes	Yes
France		No		Yes			
Germany	Yes	No	Yes	Yes	No	No	Yes
Greece	Yes	Yes	No	Yes	Yes	Yes	Yes
Hungary	No	Yes	Yes	Yes	Yes	Yes	Yes
Iceland	No	No			No	No	
Latvia	Yes		Yes				Yes
Liechtenstein	No	No	No	No	No	No	No
Lithuania	Yes	Yes	Yes	Yes	Yes	Yes	Yes
Luxembourg	Yes	Yes		Yes	No	No	Yes
Malta		Yes	Yes	No	No	Yes	
Monaco	No	No	No	No	No	No	No
Netherlands				Yes	Yes	Yes	Yes
Northern Ireland	Yes	Yes		Yes	Yes	Yes	Yes
Norway	Yes	Yes	Yes	Yes	No	Yes	Yes
Poland	Yes[b]	Yes		Yes			
Republic of Moldova	Yes	No	No	Yes	No	No	No
Romania	Yes	No	Yes	Yes	No	Yes[c]	Yes
Slovakia	Yes	No	Yes	Yes	Yes	Yes	Yes
Slovenia	Yes	Yes	Yes	Yes	Yes	Yes	Yes
Spain		Yes		Yes	Yes	Yes	Yes
Sweden	No	Yes	Yes	Yes	No	Yes	Yes
Turkey		No			No	No	

a Only epidemic cases.

b Gastroenteritis and colitis in infants from 4 weeks to 2 years of age.

c Giardiasis is included in the intestinal parasitosis rate.

often complicate the picture. In Spain, for example, records of disease incidence are based on the findings of investigations by microbiological laboratories, whereas data on diseases transmitted by water are based on notification during outbreaks. Some countries may record cases of gastroenteritis, for example, whereas others maintain records of diseases caused by individual organisms. The pathogen responsible for gastrointestinal disturbances is often not traced, and most records of these diseases do not associate the case with a cause. This is well illustrated by data from Albania, where *Shigella* spp. are known to cause 10–12% of cases of gastroenteritis, *Salmonella* spp. 2% and *E. coli* 20–25%, but the origin of the rest is unknown.

Different approaches to recording make immediate epidemiological follow-up difficult, and the exchange of information between central authorities, waterworks and local health authorities may be poor. The reports of waterborne cases of a disease often exceed the total number of laboratory-confirmed cases, leading to difficulties in interpreting the data and assessing the extent of the contribution of drinking-water to total disease incidence.

Visitors who contract diseases while on holiday contribute a significant proportion of cases of such illness in a number of European countries. Tourists are especially likely to contract enteric diseases, such as gastroenteritis, from pathogens the resident population may be able to tolerate. Other factors such as trade and societal customs may also distort the picture of disease incidence. For example, a higher incidence of bacillary dysentery in Norway in 1994 (about twice as many cases as in the previous year) was due to an outbreak caused by imported contaminated lettuce. Calculations of disease incidence based on the resident population can therefore be misleading. Statistics relating to the incidence of these diseases have to be interpreted with care.

For the period 1986–1996, surveillance data from 17 countries in the European Region (Table 5.2) reported a total of 2 567 210 cases of gastrointestinal disease, 2.0% of which were linked to drinking-water. These 17 countries (estimated population 220 million) had an average of 233 383 reported cases of gastrointestinal diseases per year. This is much lower than the estimated 6–80 million cases of

Table 5.2. Reported cases of gastrointestinal or other possibly waterborne diseases and cases of these diseases linked to drinking-water in 17 European countries,[a] 1986–1996

Causative agent and diseases[b]	Total No. (%) of cases reported		No. (%) of cases linked to drinking-water	
Bacteria: bacterial dysentery, cholera, typhoid fever and others[c]	534 732	(20.8%)	15 167	(2.8%)
Viruses: hepatitis A and Norwalk-like virus	343 305	(13.4%)	6 869	(2.0%)
Parasites: amoebic dysentery, amoebic meningoencephalitis, cryptosporidiosis and giardiasis	220 581	(8.6%)	4 568	(2.1%)
Chemicals: dental/skeletal fluorosis and methaemoglobinaemia	7 421	(0.3%)	2 802	(37.8%)
Unspecified cause: gastroenteritis and severe diarrhoea	1 461 171	(56.9%)	22 898	(1.6%)
Total	2 576 210	(100%)	52 304	(2.0%)

[a] Andorra, Austria, Croatia, Czech Republic, England and Wales, Estonia, Germany, Hungary, Latvia, Lithuania, Malta, Norway, Republic of Moldova, Romania, Slovakia, Slovenia and Sweden. On average, the countries had data available for 7 of 12 diseases (range 3–10).

[b] Information on categories/diseases in italics was not requested in the original questionnaire but was received as additional material.

[c] Others: *Aeromonas*, *Campylobacter* and *Salmonella* spp.

foodborne diseases alone in the United States (estimated population 267 million), not including, for example, cases of waterborne diseases and gastrointestinal diseases transmitted from person to person *(125,126)*. Thus the data presented here (and in the following sections) most likely underestimate the true incidence of gastrointestinal diseases in the reporting countries. This conclusion is further supported by the results of recent surveys showing that foodborne diseases may be 300–350 times more frequent than reported cases tend to indicate *(127)*; no such estimates are available for waterborne diseases.

For the 11 years from 1986 to 1996, 710 waterborne disease outbreaks were reported, an average of 3.8 outbreaks per year and country (Table 5.3). For 208 outbreaks, details about the causative agent and the type of water system were available. For 142 (68%) of the 208 waterborne disease outbreaks, the causative agent could be identified; 86 (55%) of 155 waterborne disease outbreaks occurred in rural and 69 (45%) in urban areas; 55 (36%) of 154 outbreaks were associated with networked public water supplies, 27 (18%) with individual water systems, 9 (6%) with standpipe public supplies and 63 (41%) with unspecified supplies or recreational water; and 79 (66%) of 120 outbreaks were associated with groundwater, 27 (22%) with surface water and 14 (12%) with mixed-source water.

Each outbreak for which such information was available affected an average of 220 people (range 2–3500). The wide range of waterborne disease outbreaks reported for the different countries is remarkable: during the 11-year period, none was reported in Germany, Lithuania or Norway, whereas some countries reported more than 50 such outbreaks: Spain (208), Malta (162) and Sweden (53). These differences most likely reflect not only actual differences in the incidence of waterborne disease outbreaks but also differences in detection, investigation and reporting in the different countries. For example, some countries do not have a surveillance system for waterborne disease outbreaks and do not require the reporting of waterborne diseases as such.

Although a disease outbreak may be linked to a particular source (such as drinking-water), linking a single case of (gastrointestinal) disease to a particular source is normally impossible (or impracticable). Thus, in addition to the general under-reporting of gastrointestinal diseases, the data presented here most likely underestimate by far the actual incidence of waterborne diseases in the reporting countries. Using the cases reported through surveillance systems for waterborne disease outbreaks, in order to estimate the magnitude of waterborne disease for a country or region, requires accounting for the following factors.

Table 5.3. Reported outbreaks of waterborne disease associated with drinking-water and recreational water in selected countries of the European Region, 1986–1996[a]

Country	Causative agent or disease (No. of outbreaks)	Total No. of outbreaks	No. of outbreaks for which details were available	No. of cases[b]
Albania	Amoebic dysentery (5), typhoid fever (5), cholera (4)	14	3	59
Croatia	Bacterial dysentery (14), gastroenteritis (6), hepatitis A (4), typhoid fever (4), cryptosporidiosis (1)	29[c]	31[c]	1 931
Czech Republic	Gastroenteritis (15), bacterial dysentery (2), hepatitis A (1)	18[d]	3	76
England and Wales	Cryptosporidiosis (13), gastroenteritis (6), giardiasis (1)	20	14	2 810
Estonia	Bacterial dysentery (7), hepatitis A (5)	12	12	1 010
Germany	No outbreaks reported	0	0	0
Greece	Bacterial dysentery (1), typhoid fever (1)	2	1	16
Hungary	Bacterial dysentery (17), gastroenteritis (6), salmonellosis (4)	27[e]	27	4 884
Iceland	Bacterial dysentery (1)	1	1	10
Latvia	Hepatitis A (1)	1	1	863
Lithuania	No outbreaks reported	0[f]	0	0
Malta	Gastroenteritis (152), bacterial dysentery (4), hepatitis A (4), giardiasis (1), typhoid fever (1)	162	6	19
Norway	No outbreaks reported	0	0	0
Romania	Bacterial dysentery (36), gastroenteritis (8), hepatitis A (8), cholera (3), typhoid fever (1), methaemoglobinaemia (1)	57	1	745
Slovakia	Bacterial dysentery (30), gastroenteritis (21), hepatitis A (8), typhoid fever (2)	61	61	5 173
Slovenia	Gastroenteritis (33), bacterial dysentery (8), hepatitis A (2), amoebic dysentery (1), giardiasis (1)	45	0	N/A

Table 5.3. (contd)

Country	Causative agent or disease (No. of outbreaks)	Total No. of outbreaks	No. of outbreaks for which details were available	No. of cases[b]
Spain	Gastroenteritis (97), bacterial dysentery (47), hepatitis A (28), typhoid fever (27), giardiasis (7), cryptosporidiosis (1), unspecified (1)	208	0	N/A
Sweden	Gastroenteritis (36), *Campylobacter* spp. (8), Norwalk-like virus (4), giardiasis (4), cryptosporidiosis (1), amoebic dysentery (1), *Aeromonas* spp. (1)	53[g]	47	27 074
Total	Gastroenteritis (410), bacterial dysentery (191), hepatitis A (71), typhoid fever (45), cryptosporidiosis (16), giardiasis (14), *Campylobacter* spp. (8), amoebic dysentery (7), cholera (7), Norwalk-like virus (4), salmonellosis (4), *Aeromonas* spp. (1), methaemoglobinaemia (1), unspecified (1)	710	208	44 670

[a] For the countries listed, information was available for a cumulative total of 198 surveillance years. For the period 1986–1996, Andorra, Austria, Belgium, Liechtenstein, Monaco and the Republic of Moldova had no records of waterborne disease outbreaks.

[b] N/A = not available.

[c] Discrepant data were provided in the different sections of the questionnaire.

[d] One year of reporting only.

[e] Outbreaks associated with drinking-water (n=12) and recreational water (n=15).

[f] Ten years of reporting only.

[g] In one outbreak, *Campylobacter* spp., *Cryptosporidium* spp. and *Giardia lamblia* were identified as the causative agents, and all three are therefore listed.

- The data gathered through surveillance systems for waterborne disease outbreaks probably do not reflect the actual incidence of such outbreaks, because not all of them may be recognized, investigated or reported.

- The availability and utilization of laboratory services, and the expertise of the people responsible for and the resources allocated to surveillance activities, may vary among countries.

- Recognition of waterborne disease outbreaks depends on several other characteristics, such as the severity of disease, the relative size of the outbreak and the type of water system *(128)*.

- The ratio of outbreaks to "sporadic" cases of waterborne disease is unknown and most likely varies among countries.

Hence, the data collected through the EEA/WHO questionnaire cannot be used to estimate the incidence of waterborne diseases in Europe. Nevertheless, the data imply that a large proportion of the reported gastrointestinal disease in Europe is waterborne.

Analysis of data from waterborne disease outbreaks implies that, depending on the factors mentioned previously, a high proportion of an affected population usually has to be affected before an outbreak is detected. For example, the 1993 outbreak of crypto-sporidiosis in Milwaukee, Wisconsin affected more than 400 000 people but was not detected before more than half of them had already fallen ill *(129,130)*. A retrospective epidemiological study of an outbreak of cryptosporidiosis among people with AIDS in Las Vegas in 1994 showed that nearly half the employees of two randomly selected agencies had had gastrointestinal illness during the outbreak period. This indicated that the outbreak had also affected the general population *(131,132)*, but no such outbreak was reported. These examples show that waterborne diseases break out even in countries with sophisticated water treatment facilities, and that even large waterborne disease outbreaks may not be detected.

Improved protection of source water and water treatment have markedly reduced gastrointestinal diseases in industrialized countries over the last century *(133)*. Studies and data from other parts of the world, however, suggest that much gastrointestinal illness in Europe may also still be waterborne. For example, epidemiological studies in Canada, a country with a high standard of drinking-water quality, indicate that up to 40% of reported gastrointestinal diseases (depending on the type of source water, water treatment method and distribution system) may be related to water *(134,135)*. In countries with less protected source water, less sophisticated water treatment facilities and less well maintained water distribution systems, the proportion of water-related gastrointestinal diseases is probably even higher.

The data collected by the EEA/WHO questionnaire has a number of limitations, such as differences in surveillance activities and reporting, but also limitations in data collection. The actual magnitude of waterborne diseases in Europe is therefore difficult to estimate meaningfully. Surveillance systems for waterborne disease outbreaks as well as proper networks for regulation and command *(46)* are urgently needed in the countries that do not yet have them, and the systems and surveillance definitions used in different countries need to be harmonized. The data gathered through surveillance systems for waterborne disease outbreaks will be useful in identifying the causal agents, in determining why the outbreaks occurred, in evaluating the adequacy of the water treatment technologies currently used in the different countries, and in characterizing the epidemiology of the outbreaks *(128)*. Further research is needed to be able to better estimate the burden of waterborne diseases not (obviously) related to outbreaks.

Nonspecific gastroenteritis

Gastroenteritis is an intestinal disease that can be caused by a range of microorganisms. Many countries keep records of gastroenteritis outbreaks (Table 5.1), but recording in different countries varies. Some record any incident of severe diarrhoea as gastroenteritis, while others include it in the definition of infectious intestinal diseases. Many cases of gastroenteritis are self-limiting and/or self-treated and therefore not identified by some surveillance systems.

Few countries report links of cases of gastroenteritis to drinking-water and, if they do, the cases reported are generally a small proportion of the total incidence (Table 5.2). However, this may be related to inefficient detection of outbreaks. Nevertheless, a number of countries in the European Region regard waterborne gastroenteritis as a serious problem.

Amoebic dysentery

Ameobic dysentery is a debilitating disease caused by the protozoan *Entamoeba histolytica*. The symptoms include abdominal pain, diarrhoea alternating with constipation, or chronic dysentery with discharge of mucus or blood. Carriers of amoebic dysentery are now found worldwide. It is believed to be carried by only a small proportion of the population of Europe. Cysts of *E. histolytica*, like all protozoan cysts, tend not to settle in wastewater treatment plants, and sewage effluent may therefore contaminate surface water *(136)*.

The number of cases of amoebic dysentery reported from countries that maintain records is generally low (Table 5.1 and Fig. 5.3), although between 1000 and 4000 cases were reported annually in Sweden in the early 1990s (an incidence of 11–45 per 100 000 population per year). Slovenia has had the only reported cases of amoebic dysentery known to be linked to drinking-water; 39 of a total 46 cases in 1991 were believed to have resulted from contaminated drinking-water.

Fig. 5.3. Reported incidence of amoebic dysentery in selected European countries, 1996

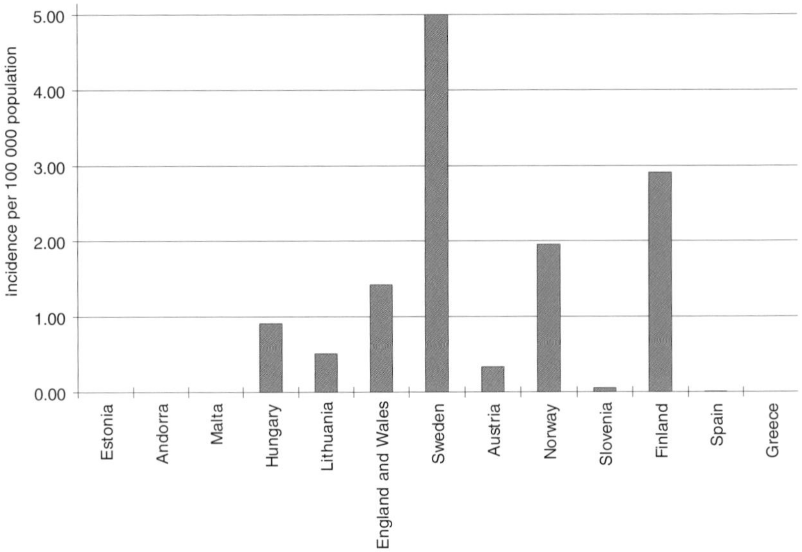

Note: Most of the cases in Finland were contracted abroad.

Bacillary dysentery

Bacillary dysentery is an infectious intestinal disease caused by *Shigella* spp. The disease is spread primarily by person-to-person contact, as *Shigella* spp. rarely infect animals and do not survive well in the environment. The infectious dose is low. Poor-quality drinking-water contaminated by sewage has caused disease outbreaks *(136)*, although chlorination readily destroys *Shigella (105)*. About half the countries in Europe keep records of bacillary dysentery (Table 5.1). Outbreaks are regularly reported in many countries (Fig. 5.4 and Table 5.2). In an epidemic of *Shigella sonnei* gastroenteritis in Israel, thought to be waterborne, 1216 people were affected within 3 weeks, 302 of them members of communal settlements, the kibbutzim. People at high risk within the kibbutzim were temporary visitors from Europe and the United States, children aged 1–5 years, adult women, and children and their mothers in kibbutzim *(137)*.

Fig. 5.4. Reported incidence of bacillary dysentery in selected European countries and regions, 1996

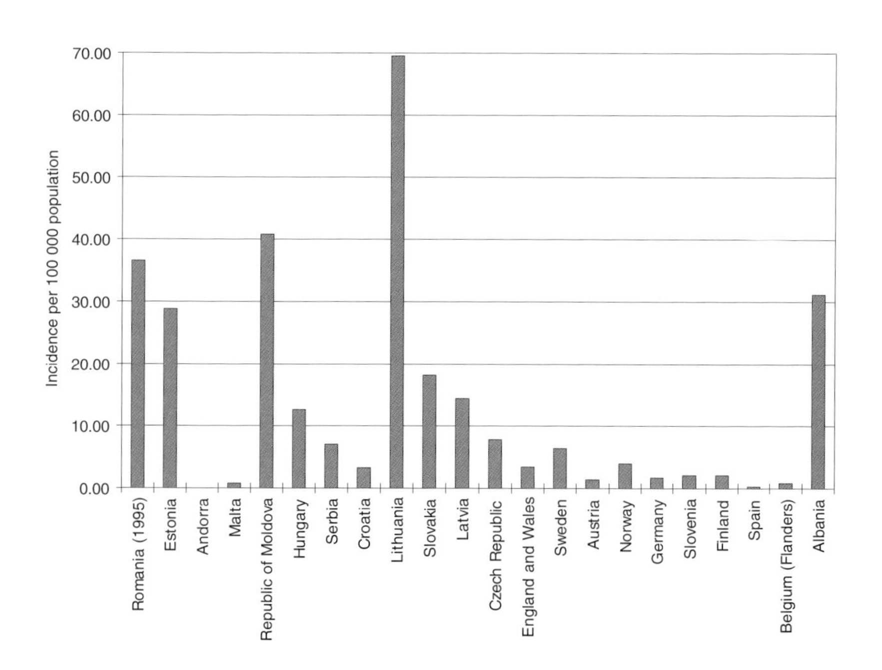

The contribution of known waterborne cases to total morbidity varies between the countries for which data are available and between years, but it is significant in some cases. In Spain between 1986 and 1995, a number of cases (between 83 and 1327) annually were linked to drinking-water. A number of countries, including Albania, Greece and Spain, regard bacillary dysentery as one of the most serious waterborne disease problems.

Campylobacteriosis

Campylobacter spp. are spiral bacteria that cause severe acute diarrhoea. They have been isolated from surface water contaminated by sewage from farm animals and wildlife, especially birds, although they are susceptible to chlorination *(105)*. However, the most important reservoirs of the bacterium are meat and unpasteurized milk. Household pets can also carry the bacterium. *Campylobacter* can remain viable for extended periods in the environment, particularly at relatively low temperatures, and survival in excess of 12 months is possible at 4 °C *(105,136)*. Six waterborne outbreaks were recorded in Sweden between 1986 and 1996. *Campylobacter* has been reported to be widespread in the River Moskva and its tributaries in the newly independent states *(23)*.

Cholera

The symptoms of cholera are sudden diarrhoea with watery faeces, accompanied by vomiting, and the resulting dehydration and collapse are fatal in over half of untreated cases. The main routes of transmission are waterborne or foodborne. Direct person-to-person contact is uncommon.

Vibrio cholerae causes cholera. There are about 139 known O serotypes, of which only two are known to be responsible for epidemics of cholera (serotypes O1 and O139) *(138)*.

The epidemiology of cholera is characterized by its tendency to spread throughout the world in pandemics. The first six pandemics began in Bangladesh. The seventh started in 1961 in Indonesia and spread to the Indian subcontinent, the USSR, Iran and Iraq during the 1960s *(138)*. In the USSR, 11 republics reported 10 723 cholera cases and carriers between 1965 and 1989. Since then the epidemiological situation in the newly independent states has been unstable, and cases of cholera and the isolation of virulent stains from surface water are reported every year *(139)*. Cholera epidemics have re-emerged since 1991, both in the continents where the disease is epidemic and in traditionally cholera-free areas *(140)*, although the number of reported cases in most European countries is low. Cholera has basically been controlled in Europe by improved water and sewage treatment and improved food hygiene, although nontoxigenic *V. cholerae* is not uncommon in European surface water *(141)*. All pathogenic *Vibrio* spp. are halophilic and survive better in moderately saline water. Temperatures of at least 10 °C for several consecutive weeks are also important for the survival of *V. cholerae* in the environment. The organism is highly susceptible to chlorine and is readily eliminated from water by proper disinfection.

Tourism by western Europeans in the western Pacific, South-East Asia and other areas is likely to have led to the significant increase in the proportion of those populations carrying the organism. All cases reported from Andorra, England and Wales, Finland, Greece and Sweden between 1986 and 1996 were in individuals who had contracted the disease abroad. Most cases recorded in Spain are also imported. However, several countries in central and eastern Europe appear to have domestic cases. In Romania, where records on cholera as a notifiable disease have been kept since 1986, drinking-water caused an estimated 286 cases between 1991 and 1993. These occurred in three outbreaks, one each year, and all occurred within the Danube delta. In Albania, 626 cases of cholera were reported in 1994, 25 of which were fatal. All of these were linked to drinking-water, with four outbreaks of waterborne cholera reported in that year. The incidence of cases of cholera in the Republic of Moldova rose from 1 in 1991 to 240 in 1995. Although none was reported to be linked to drinking-water, cases were concentrated in particular areas (Slobozia, Stefan Voda and Tiraspol districts). A cholera epidemic in Ukraine in 1994 and 1995, with 1370 recorded cases, resulted in 32 fatalities. The environmental sources of the causative agent included sewage, seawater and surface water.

Cryptosporidiosis

Cryptosporidium is a coccidial protozoan parasite. About 20 species are now known, of which *C. parvum* is pathogenic for humans.

Infection causes gastroenteritis with stomach cramps, nausea, dehydration and headaches. The disease is usually self-limiting and lasts for up to two weeks, but can be fatal in very young and very old people and in those who are immunosuppressed, such as those with AIDS *(137)*. *Cryptosporidium* is widespread in nature and has a wide range of animal hosts in addition to humans. In its protected stage (the oocyst), it is able to survive for several months in water at 4 °C *(142)*. Treatments conventionally used to remove particulate matter and microbial contamination from surface water, such as coagulation and filtration, remove a large proportion of the oocysts. Because of their small size (4–7 μm in diameter) and their resistance to disinfection, however, oocysts are difficult to remove by water treatment *(122)*. Exposure to a small number of oocysts is believed to be sufficient to cause infection. A number of European countries keep records of detected cases of cryptosporidiosis (Table 5.1).

C. parvum infection occurs worldwide in urban and rural populations *(143)*. Several thousand cases were recorded each year from 1986 to 1996 in England and Wales (Table 5.4) *(143)*. Comprehensive records are not available for those cases associated with contaminated drinking-water, but some outbreaks have been investigated and are believed to have originated from that source. For the period 1986–1996, 13 outbreaks of waterborne cryptosporidiosis were recorded in England and Wales. About 4% of all cases of cryptosporidiosis reported in the United Kingdom were in people who had recently returned from abroad.

At least 500 cases of cryptosporidiosis were confirmed in outbreaks in England and Wales between 1989 and 1995, with an additional 5000 people possibly being affected in the 1989 incident. Although smaller numbers than this (between 20 and 477) were involved in other possible waterborne outbreaks, these incidents demonstrate the

Table 5.4. Number of cases of cryptosporidiosis reported in England and Wales, 1986–1996

	Year										
	1986	**1987**	**1988**	**1989**	**1990**	**1991**	**1992**	**1993**	**1994**	**1995**	**1996**
No. of cases	3565	3277	2750	7768	4682	5165	5211	4832	4433	5684	3662

potential for contaminated supplies to cause the disease in a significant number of people. All outbreaks in the United Kingdom detailed in the EEA/WHO questionnaire were in networked public supplies.

Cryptosporidium oocysts have been found to be widespread in water resources in Great Britain, and they have been shown to be present at higher concentrations in lakes and rivers receiving wastewater than in pristine streams *(136)*. The source of environmental contamination can be both human sewage and animals. Most probable or possible waterborne outbreaks in England and Wales have been associated with supplies derived from surface-water sources, as might be expected given their greater likelihood of contamination by run-off from agricultural land, wildlife faeces and sewage discharges. The seasonal presence of the organisms in the source water supplying the treatment plant involved in the 1989 incident was particularly associated with the grazing of young lambs *(136)*. Nevertheless, at least three outbreaks have been linked to groundwater sources.

Investigations into a number of waterborne outbreaks have established that the water supplied was free of indicators of faecal contamination. This emphasizes the resilience of the oocysts to disinfection. Water suppliers face difficulties in detecting and efficiently removing *Cryptosporidium (121,136,144)*. Recycling of filter backwash water has also been shown to contribute to the build-up of oocysts in wastewater treatment plants. Proper treatment of washwater before recycling may prevent such a build-up *(131)*. Recreational water use, contact with farm animals and old piping have also been identified as risk factors for cryptosporidiosis *(145,146)*.

Giardiasis

Giardia lamblia is found in its free-living form in a wide range of host animals. Its cysts can survive in unchlorinated water for long periods, especially in cold weather. Symptoms of giardiasis develop in the first few weeks after infection and include severe, watery, foul-smelling diarrhoea, gas in the stomach or intestines, nausea and loss of appetite. Water is probably not the primary mode of transmission of giardiasis, but it is a common transmitter *(124,136)*.

The reported incidence of giardiasis in European countries varies enormously, and people returning from travel abroad probably import a significant proportion of cases in some countries. One known incident in the United Kingdom involved a community-managed groundwater supply in a rural area supplying 260 people. Of these, 31 (12%) contracted the disease.

In countries where giardiasis has been known to be spread through water supplies, the contribution of waterborne infection to the total disease burden varies significantly between years. In Sweden, for example, 3500 waterborne cases were recorded in 1986. This was an exceptional outbreak caused by an overflow of sewage into the drinking-water system at a ski resort *(147)*. Between 1990 and 1996, more than 23 000 cases were reported in Sweden. Known waterborne infections comprised less than 1% of infections in 1990 and 1996; this figure was as high as 14% in 1991, although no cases were linked to drinking-water in most years of the 10 years up to 1996.

Similar variation was reported in Slovenia, where the number of cases of giardiasis reported annually varied between 329 and 1299 in the 10 years up to 1996. In 1992, 40% of these cases (520 of 1299) were linked to drinking-water. Several cases in Spain have also been linked to drinking-water.

Helminthiasis

The helminths, or parasitic worms, belong to two unrelated groups of organism: roundworms and flatworms (flukes and tapeworms). Their distribution is limited geographically, and few of significance to human health are found in Europe. In sub-Saharan Africa, drinking-water is a primary mode of transmission only for the guinea worm (*Dracunculus medinensis*), a roundworm whose distribution is limited to certain countries. Other helminths may also be transmitted by drinking-water, but this is unlikely to be the most important route of infection *(105)*.

Schistosoma spp. can affect the intestine, causing intestinal schistosomiasis, or the blood vessels around the bladder, causing urinary schistosomiasis. Organisms causing both forms of the disease occur in the Eastern Mediterranean Region of WHO but are not found in the European Region. The infective larvae are able to penetrate the human skin or mucous membranes, and contaminated water used for washing, irrigation or recreation is the main hazard rather than drinking-water *(105)*.

Other helminths that could potentially be transmitted through drinking-water include *Fasciola* spp. (flukes), which are parasites of farm and domestic animals. The eggs of the pork tapeworm (*Taenia solium*) may survive in the environment after they are excreted in faeces, and will infect humans if ingested. The usual route of infection is the consumption of undercooked pork *(105)*.

Infection by waterborne helminths is not a significant risk in most parts of the European Region, although it is of potential concern following the use of wastewater in agricultural irrigation. Epidemiological studies have shown that crop irrigation with untreated wastewater causes a significant increase in intestinal nematode infections in consumers and field workers, but that this is not the case when wastewater is adequately treated before being used for irrigation. WHO guidelines for the microbial quality of treated wastewater intended for crop irrigation recommend that the treated wastewater contain less than one viable intestinal nematode egg per litre. Wastewater complying with this guideline will contain few, if any, protozoan cysts, and no (or, exceptionally, very few) *T. solium* eggs, so consumers and field workers will also be protected from protozoan and tapeworm infections *(89)*.

Infectious hepatitis

Viral hepatitis is an inflammation of the liver caused by one of several different viruses. Two of these, hepatitis virus A and hepatitis virus E, have been regularly associated with waterborne outbreaks of disease. Hepatitis virus A is common throughout the world and is very infectious. It causes nausea, vomiting, muscle ache and jaundice and is spread by faecal contamination of food, drinking-water or water used for bathing and swimming. Hepatitis virus E is less common and is restricted to tropical and subtropical countries.

Most countries keep records of reported cases of hepatitis A (Table 5.1). The number of cases varies enormously between countries, with the incidence in the central Asian republics and other newly independent states being particularly high. The incidence in the EU, Norway and Iceland is much lower (Fig. 5.5).

Some cases have been linked to contaminated drinking-water (Table 5.2). The extent to which the origin of a case is traced is also

Fig. 5.5. Incidence of hepatitis A per 100 000 population in the WHO European Region, 1996

Source: Health for all database, WHO Regional Office for Europe.

likely to vary considerably between countries. For example, the incidence of infections in England and Wales fluctuated substantially between 1986 and 1997. The proportion considered to have been acquired abroad fluctuated also and was estimated to be between 3% and 20% for this period.

Norwalk and Norwalk-like viruses

Norwalk virus causes severe diarrhoea and vomiting. Several waterborne outbreaks caused by Norwalk or Norwalk-like viruses have occurred in Norway and Sweden, and both countries regard it as a serious problem. In Sweden, four outbreaks occurred between 1994 and 1996, in which 325 people contracted the disease. It is currently unclear whether normal chlorination practice kills Norwalk virus *(105,136)*.

Non-gastrointestinal disease – typhoid fever

Salmonella spp. cause typhoid fever (*Salmonella typhi*) and paratyphoid fever (*Salmonella paratyphi*). The current incidence of typhoid fever is low in most European countries that keep records (Table 5.1, Fig. 5.6) and, as with cholera, a significant proportion of cases recorded in many countries are likely to have been imported. All cases reported in Flanders (Belgium) in the 10 years up to 1996 were contracted abroad, and this is also true of almost all of the cases reported in Finland, Norway and Sweden. About 90% of cases reported in the United Kingdom were acquired outside the country.

Some countries (Albania, Croatia, Estonia, Greece, Romania, Slovakia and Spain) have linked recorded cases of typhoid fever to drinking-water supplies during the last decade. In Romania, the incidence of typhoid dropped steadily from 1986 (0.3 per 100 000 population to 1995 (0.04 per 100 000 population) and only nine cases were linked to drinking-water.

Fig. 5.6. Reported incidence of typhoid fever in selected European countries and regions, 1996

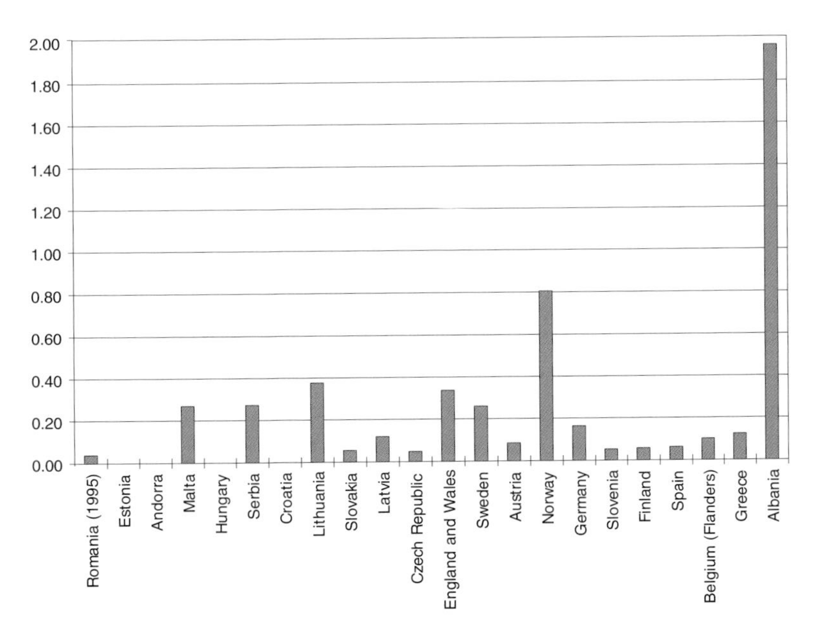

Tajikistan had an outbreak of typhoid fever (identification based on clinical observations) in 1996. This was the result of heavy rainfall causing overflow from the poorly maintained sewerage systems, with subsequent contamination of drinking-water sources. Nearly 4000 cases were reported between mid-May and the beginning of July, and the outbreaks were still continuing. Ten districts in two provinces were affected and a number of deaths occurred, mainly among people between 10 and 29 years of age. The fatality rates reported in the two provinces were 1.2% and 8.2% *(107)*.

HEALTH PROBLEMS ASSOCIATED WITH WATERBORNE CHEMICALS

Lead

Lead is a cumulative, general toxic substance that can affect a large number of processes in the body. Exposure to relatively low levels of lead impairs the development of the nervous system in children, leading to reduced performance on standardized intelligence tests. This is the effect of greatest concern for low-level exposure. Effects on gestational age and fetal weight are also important, as are blood pressure and some biochemical effects. Infants, children up to 6 years of age, the fetus and pregnant women are the most susceptible to the adverse health effects of lead. Endemic diseases of the nervous system, gingivitis and hypermenorrhoea have been observed in the Debet River basin in Armenia, where concentrations of lead in freshwater ecosystems are 5–200 times higher than normal *(148)*.

Concerns about the potential health effects of dissolved lead have resulted in considerable efforts to reduce lead concentrations in water. Nevertheless, lead piping is retained in some properties. It is this pipework, owned by and the responsibility of the householder, that has been responsible for most of the failures of drinking-water samples to meet the standard of the EU Drinking Water Directive (until 1998) of 50 µg/l *(47)* and the WHO guideline value of 10 µg/l *(105)*. In the years 1990–1995, 2.7–3.4% of drinking-water samples tested in England and Wales exceeded 50 µg/l. The revised Drinking Water Directive that entered into force on 25 December 1998 *(48)* adopted a new standard identical to the WHO guideline value of 10 µg/l, and a greater proportion of samples are therefore likely to exceed this standard. Nevertheless, the EU countries do not have to fully implement the new standard until 2013.

Several water suppliers in the United Kingdom, especially those distributing water with a low mineral content, use orthophosphate dosing to reduce the dissolution of lead (Box 5.2).

Concern in Hungary has been caused by lead leaching out of newly laid polyvinyl chloride pipes, in which it had been used as a stabilizer. Such leaching appears to be a short-term problem, however, with concentrations falling to below the national standard of 50 μg/l a few weeks after the pipes are first used.

Arsenic

Inorganic arsenic is a known human carcinogen and is classified by IARC in Group 1 (a known human carcinogen). The WHO provisional guideline for arsenic is 0.01 mg/l *(101)* and this has been adopted as the standard in a number of European countries, although the standard in most remains at the previous WHO provisional guideline value of 0.05 mg/l *(149)*. The guideline value of 0.01 mg/l is provisional because suitable testing methods are lacking. Based on health concerns alone, the guideline would be even lower. Data from Belgium, Croatia, Greece, Monaco, Slovakia and the United Kingdom (England, Wales and Northern Ireland) indicate that arsenic in drinking-water does not generally exceed the national standard of 0.05 mg/l in these countries. In France, only 0.06% of samples tested in 1995 exceeded this standard, and the figure for the Czech Republic in 1996 was 0.25%.

Box 5.2. Treatment to reduce the concentration of lead in drinking-water supplies in the United Kingdom

Between 1994 and 1996, several areas of England and Wales undertook to reduce the leaching of lead into drinking-water, usually by orthophosphate dosing or by adjusting the pH. In 1996, 2.2% of drinking-water samples tested exceeded the national standard, compared with 2.7–3.4% between 1990 and 1995. Concentrations of lead in drinking-water are also a concern in Scotland, where the predominant soft, acidic water readily dissolves lead. Several educational drives have been mounted to minimize the exposure to lead of children living in housing with lead piping, for example by advising householders to avoid using water that has been standing in the pipework for several hours.

Consumption of drinking-water containing high levels of arsenic has been associated with increased skin cancer and possibly internal types of cancer *(105)*. Concentrations exceeding the WHO guideline value are found in some countries in central Europe, especially where there are natural sources of arsenic in minerals (Box 5.3). In south-eastern Hungary and in the adjoining areas of Romania, especially in individual and public wells in Bihor and Arad Counties, 3000 people were exposed to levels exceeding the national standard (Institute of Public Health, Cluj-Napoca). In Bulgaria, high concentrations have been found in both surface and well water near a copper smelter at Srednorgie. The south-western parts of Poland have naturally high levels of arsenic, although concentrations in drinking-water do not generally exceed the national standard.

Arsenic is not routinely monitored in all countries. Sweden, for example, does not routinely monitor for arsenic, and 17% of the samples taken during two surveys in 1986 exceeded the national standard of 0.01 mg/l. These were in water from private wells. Nevertheless, the percentage for Sweden as a whole is much lower; arsenic causes an estimated 0.3–3 cases of cancer per year *(103)*. Some areas of Finland have naturally elevated arsenic concentrations in the groundwater: local drilled wells serving fewer than 50 people and private drilled wells for single households.

Fluorosis

Fluorides exist naturally in a number of minerals; the most common are fluorspar, cryolite and fluorapatite *(105)*. Where the geology is rich in such minerals, groundwater sources may contain high levels of fluoride, with levels of up to 10 mg/l in well water *(105)*. Fluoride can enter surface water as a result of industrial discharge, although the levels are usually lower than the highest that can be found in groundwater. The section of the River Meuse in France, for example, has concentrations fluctuating between 0.2 mg/l and 1.3 mg/l as a result of variation in industrial processes *(105)*. Ingesting high levels of fluoride can cause mottling of developing teeth; such dental fluorosis does not affect the strength of teeth, however, and is often regarded as a cosmetic rather than a toxic effect. The more serious toxic effect of excessive fluoride intake is skeletal fluorosis, in which the size of the bone mass increases. This restricts movement, both mechanically and because nerves that pass through bones are constricted. In severe cases, paralysis can result.

Box 5.3. Reducing arsenic levels in drinking-water in Hungary

Groundwater in some areas of Hungary contains high natural concentrations of arsenic. This problem came to light in 1981, when a countrywide survey indicated that more than 400 000 people were exposed to drinking-water containing arsenic at levels above the national standard of 0.05 mg/l. Measures were rapidly taken to address the situation and to reduce the number of people exposed to concentrations above the health-based standard.

A number of different approaches have been used, depending on the size of the supply and the properties of the source water. In some smaller villages and in a bigger plant (10 000 m^3 per day) serving two towns, co-precipitation technology was used to remove the arsenic. Trivalent arsenic, As(III), is oxidized to the pentavalent form, As(V), usually by chlorine. Iron(III) salts are dosed and the pH corrected using lime, to coagulate and precipitate the arsenic, and the water filtered to remove the precipitated salts. Where the water also contains humic material, a second oxidation and flocculation was undertaken using permanganate, and the water refiltered.

A simpler solution is possible where the natural concentration of arsenic is not much higher than the standard (such as 0.060–0.075 mg/l) and the water also has a natural iron content of 0.6–0.7 mg/l. Only pre-oxidation using chlorination is required to induce co-precipitation prior to filtration, and this treatment has the added benefit of removing the iron, which would otherwise have required aeration and filtration. This procedure has been used in several Hungarian villages.

Where possible, such treatments are avoided by using an alternative water source to supply the population. New abstractions have been established to supply water from aquifers some distance from the settlements. A new regional waterworks has been established to supply water from this alternative source to about 20 settlements, including a large town of some 80 000 inhabitants. The majority of the older, contaminated wells have remained in operation and, in most cases, this water is mixed with the uncontaminated water to produce water containing a reduced level of arsenic.

These solutions were successful in reducing the concentrations of arsenic in drinking-water supplies to below the old standard of 0.05 mg/l. The number of people estimated to be exposed to concentrations exceeding this level had been reduced from 400 000 to 20 000 by the end of 1995. However, Hungary has recently revised the standard for arsenic in drinking-water to 0.01 mg/l, in line with the most recent provisional WHO guideline *(105)*, and the solutions currently used to reduce arsenic levels are not adequate to ensure compliance with this level. A new countrywide survey, begun in 1997, is under way to establish how many people may be exposed to concentrations higher than this. The preliminary results suggest that, although less than 0.5% of the population is exposed to arsenic concentrations higher than 0.05 mg/l, 10–12% may receive drinking-water with a concentration higher than 0.01 mg/l.

Source: M. Csanady, personal communication, 1998.

Fluorine is probably an essential element for humans but this has not been demonstrated unequivocally, and no data indicating the minimum nutritional requirements are available. Many epidemiological studies of possible adverse effects of the long-term ingestion of fluoride via drinking-water have been carried out and clearly establish that fluoride primarily affects skeletal tissue. Nevertheless, low concentrations provide protection against dental caries, especially in children. This protective effect increases with concentrations up to about 2 mg/l; the minimum concentration required is about 0.5 mg/l *(105)*. Some countries and cities add fluoride to drinking-water at low levels to promote dental health. In these cases, a final concentration of about 1 mg/l is generally considered desirable. Populations in south-eastern Slovenia, where the concentrations are < 0.8 mg/l, report a high rate of dental caries.

The WHO maximum guideline value is 1.5 mg/l *(105)*, and most European countries have adopted this or a similar figure (Hungary 1.7, Ireland 1.0, Italy 1.5–0.7,[1] the Netherlands 1.1, Romania 1.2 and Sweden 1.3 mg/l) as the standard for fluoride in drinking-water. In Poland the recommended fluoride concentration is 0.3 mg/l or higher.

Most countries do not keep records of fluorosis. In France, 1.3% of samples from supplies covering over 5000 people exceeded the national standard in recent years, but in all these cases the level was below 4.5 mg/l. In other countries, a significant proportion of drinking-water samples contain fluoride at levels exceeding the national standard, although the estimated population exposed may be relatively small (Fig. 5.7). In Estonia, for example, 25–35% of the drinking-water samples analysed for fluoride since 1988 contained levels exceeding the national standard, but the number of people exposed to these high levels was estimated to be 0.7% of the total population. Incidents of dental fluorosis in Sweden are scattered around the country, depending on the underlying geology, and an estimated 2.4% of the population is affected.

About 50 000 wells in Sweden are estimated to have levels of fluoride higher than 1.3 mg/l, and 1200 wells exceed 6 mg/l *(103)*. Skeletal fluorosis is very rare, but a few cases have occurred near glassworks, where wells have become contaminated with fluorine-based acid used for etching glass. An estimated 35% of the population of the Republic of Moldova is exposed to drinking-water containing fluoride at concentrations above the national standard.

[1] Temperature-dependent: 1.5 mg/l at 8–12 °C and 0.7 mg/l at 25–30 °C.

Fig. 5.7. Percentage of drinking-water samples in selected European countries exceeding the national standard for fluoride in the years specified[a]

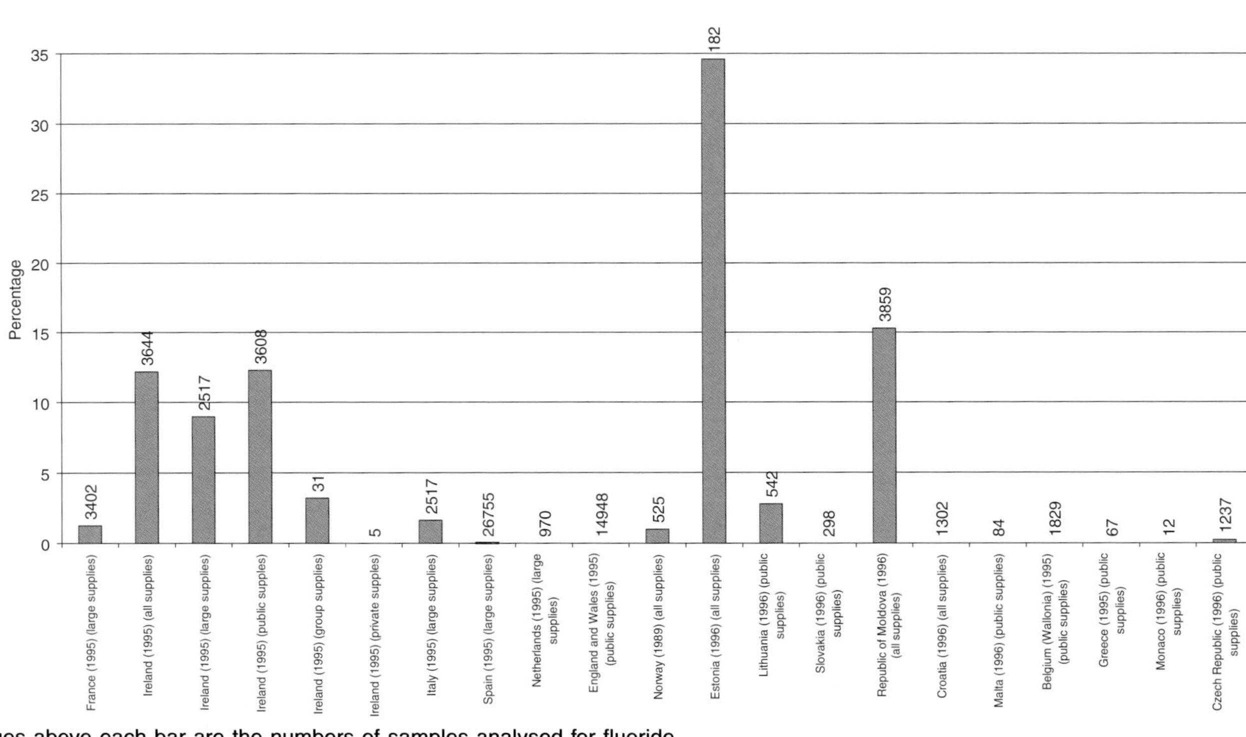

[a] Values above each bar are the numbers of samples analysed for fluoride.

In the Czech Republic, an estimated 1600 cases of dental fluorosis were reported annually between 1991 and 1993 in parts of central and northern Bohemia, of which 100 annually were linked to drinking-water. The area receives water from the central Bohemian chalk. About 30 cases per year of skeletal fluorosis were recorded between 1991 and 1993. Although these were not directly linked to drinking-water, they are concentrated in the same area.

Fluoridation is widely practised in Ireland. The national standard for fluoride in Ireland is reported to be exceeded, as the addition of fluoride to public water supplies is poorly controlled. None of the five samples from community-managed supplies exceeded the standard in 1995, and fewer group supplies (3.2%) exceeded the standard than public supplies (12.3%). The percentage of samples exceeding the standard for fluoride in Ireland has dropped from 15.9% to 12.2% (all supplies), as many authorities are addressing shortcomings in the process of adding the fluoride during the water treatment process *(120)*.

Cyanobacterial toxins

The eutrophication of surface water often results in blooms of algae in the surface layers. Cyanobacteria (blue-green algae) threaten the quality of drinking-water by causing physical effects on drinking-water treatment equipment, and some blooms contain species that can produce toxins *(37)*.

Blooms of toxin-producing algae are a particular hazard to users of recreational water. The toxins are normally retained within the algal cells and normally present no immediate danger to drinking-water supplies, since the algal scum is readily removed by filters. If the bloom begins to break down or the filters become blocked, however, the toxin may be released directly into the water. Conventional coagulation treatment does not remove cyanobacterial toxins. Advanced treatments such as powdered or granulated active carbon, ozonation, potassium permanganate or nanofiltration can be effective in removing or breaking down cyanobacteria *(37,150–152)*.

Several toxic compounds produced by these algae have been identified and characterized according to their effects. The most commonly encountered toxins are hepatotoxins. The most potent of these is a cyclic septa-peptide termed microcystin-LR, which on acute exposure has been shown to cause the death of laboratory animals from massive hepatic haemorrhage. More than 40 variants of this toxin have been discovered, and a related hepatotoxic cyclic peptide, nodularin, has also been identified. Microcystin-LR has not been adequately investigated as a chronic toxin, but there is evidence that it can produce cellular changes that could promote tumour formation *(153)*.

The principal neurotoxin identified is anatoxin-a, which is up to two orders of magnitude more potent than nicotine. This appears to be of concern only as an acute toxin, whereby it can induce paralysis of respiratory muscles. Two others, saxitoxin and anatoxin-a(s) have also been identified; these inhibit nervous system function. Anatoxin-a(s) is unstable in water.

A third class of toxic compound associated with some algal blooms has also been identified. Some lipopolysaccharides are capable of causing skin disorders, including irritation, rashes and wheals. Gastrointestinal effects have also been recorded.

Authorities in Sweden are very concerned about the presence of algal toxins in drinking-water, as progressive eutrophication has allowed dense algal blooms to develop in surface water almost annually. Similar problems have been reported in many countries worldwide *(154)*, including many in Europe *(155–157)*.

Methaemoglobinaemia, nitrate and nitrite

Several European countries report high nitrate concentrations in drinking-water, including Austria, Belgium, Croatia, the Czech Republic, England and Wales, Estonia, France, Germany, Malta, the Netherlands, the Republic of Moldova, Slovakia, Slovenia, Turkey and Ukraine. In France, an estimated 3.5% of the population is exposed to concentrations of nitrate between 50 and 150 mg of N per litre in drinking-water *(50)*. Fig. 5.8 shows the percentage of drinking-water samples exceeding the national standard for nitrate in some European countries in 1995.

High concentrations of nitrate in drinking-water are of concern because nitrate can be reduced to nitrite, which can cause

Fig. 5.8. Percentage of drinking-water samples exceeding the standard for nitrate (WHO standard of 50 mg/l or lower national standard) in 1995 in selected European countries[a]

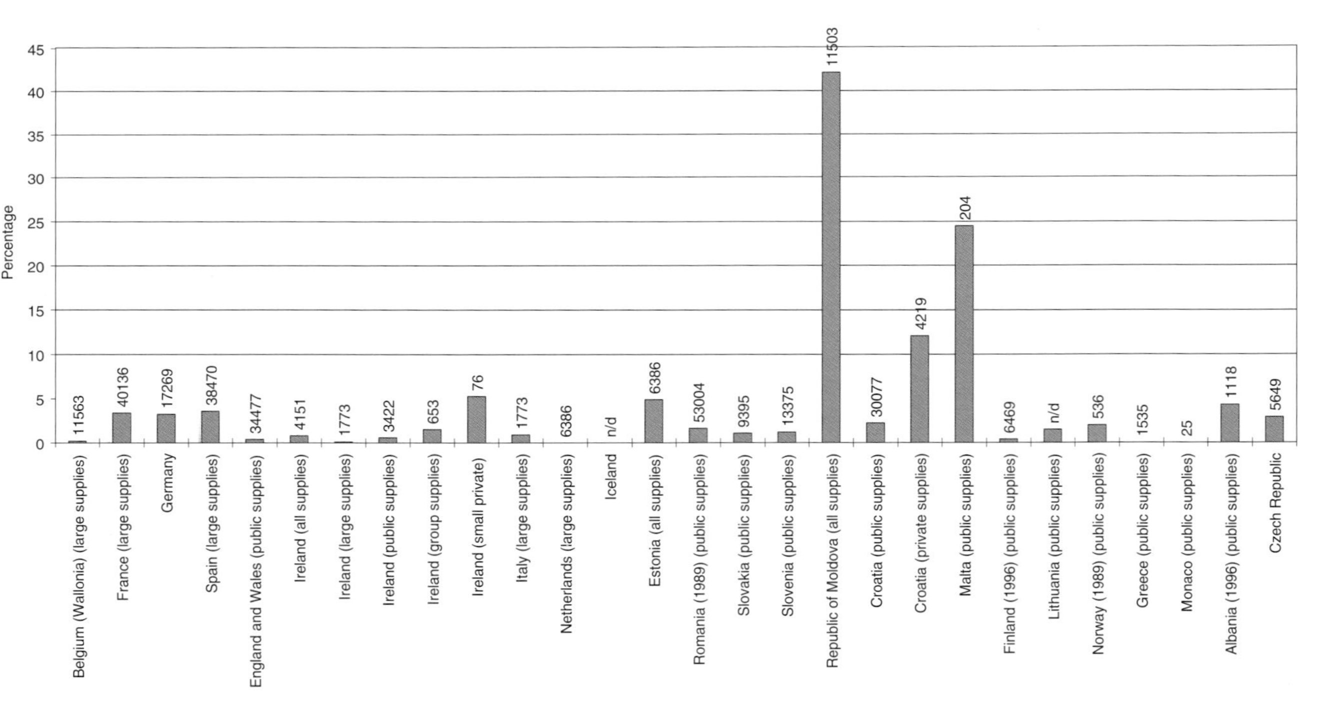

[a] Values above each bar are the numbers of samples analysed for nitrate.

methaemoglobinaemia. The haemoglobin of young children is particularly susceptible to methaemoglobinaemia and this, coupled with the increased ratio of water consumption to body weight, makes infants especially vulnerable to the disorder. When methaemoglobin levels rise to about 10% of the total haemoglobin, the so-called blue-baby syndrome develops. Progressive symptoms are stupor, coma and, in some cases, death.

Methaemoglobinaemia is unusual in older children, even where nitrate levels are quite high. Infants fed on breast-milk are not at risk, unlike those bottle-fed on feed made using nitrate-rich water. Nevertheless, populations supplied by properly disinfected drinking-water have had few, if any, cases of methaemoglobinaemia. An infant with a bacterial infection, such as gastroenteritis, is more likely to suffer from methaemoglobinaemia than is a healthy child. It is thought that this is because the enhanced bacterial activity associated with an infection increases the reduction of nitrate to nitrite. Few countries keep records of methaemoglobinaemia (Table 5.5), and incidents are few (Fig. 5.9). Most reported cases of methaemoglobinaemia associated with drinking-water are caused by well water or community-managed water supplies of poor microbial quality *(105)*.

Groundwater sources, especially those fed by percolation from intensively farmed agricultural land, are liable to be contaminated by nitrate. The vulnerability of infants receiving drinking-water from shallow groundwater sources is emphasized by data from Hungary, where between 9 and 41 cases of methaemoglobinaemia associated with drinking-water are reported annually. All cases are related to individual private wells, and almost the whole country is affected except for the south-eastern part, where deep well water is used. A similar number of water-related incidents are recorded annually in Slovakia and, as in Hungary, most of these are associated with drinking-water. In both Hungary and Slovakia, more than 80% of the recorded cases of methaemoglobinaemia are reported to be linked to drinking-water. In Albania, all 43 cases reported in 1996 were linked to nitrate in drinking-water.

The vulnerability of rural populations supplied by community-managed supplies to high concentrations of nitrates is illustrated by data from Romania (Box 5.4) and Lithuania (Fig. 5.10 and 5.11). In

Table 5.5. Countries in the European Region reported to be keeping records on methaemoglobinaemia

	Keep records	Do not keep records
Albania	✓	
Andorra		✓
Austria		✓
Belgium		✓
Croatia		✓
Czech Republic		✓
England and Wales		✓
Estonia		✓
Finland	✓	
France		✓
Germany		✓
Hungary	✓	
Latvia		✓
Liechtenstein		✓
Lithuania		✓
Luxembourg		✓
Malta		✓
Monaco		✓
Netherlands		✓
Northern Ireland	✓	
Norway		✓
Republic of Moldova		✓
Romania	✓	
Slovakia	✓	
Slovenia	✓	
Sweden	✓	

Lithuania in the last decade, less than 1.5% of samples taken from public water supplies sourced from groundwater exceeded the national standard for nitrate (50 mg/l). In contrast, in 1989 nearly 50% of samples from private water supplies sourced from groundwater contained concentrations of nitrate exceeding the standard. The consumers of some of these supplies were exposed to water containing nitrate concentrations of over 150 mg/l, considerably higher than the WHO guideline for nitrate in drinking-water of 50 mg/l *(105)*.

The vulnerability of private supplies is also illustrated by the situation in Sweden, where an estimated 35 000 people (0.4% of the population) provided with drinking-water from wells in southern Sweden

Fig. 5.9. Incidence of methaemoglobinaemia in selected European countries, 1996

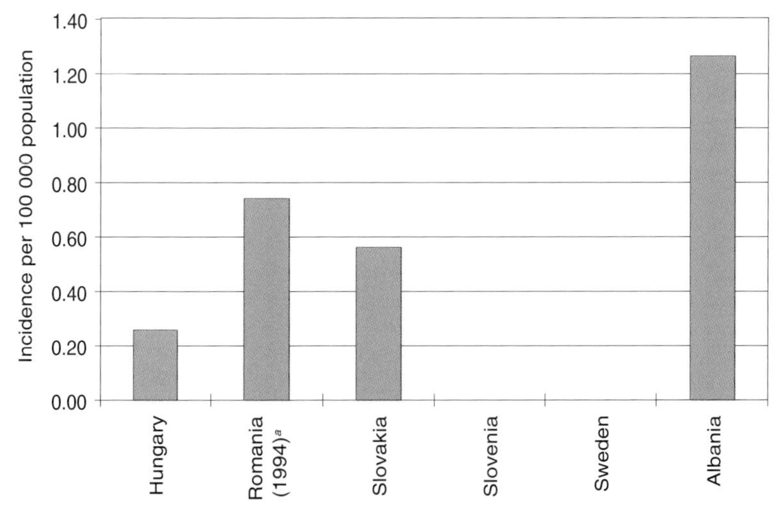

[a] Records for methaemoglobinaemia in Romania are for cases related to well water only.

Box 5.4. Nitrate in wells in Romania and methaemoglobinaemia

Methaemoglobinaemia has been recognized as a problem in the rural areas of Romania since 1955, and records have been kept of cases related to well water since 1984. Between 1985 and 1996, 2913 cases were recorded, of which 102 were fatal. Analysis of the data on 954 cases of infantile methaemoglobinaemia occurring between 1990 and 1993, of which 37 were fatal, indicated that 383 were bottle-fed, 332 were both breast-fed and bottle-fed and only 239 were breastfed.

Methaemoglobinaemia was associated with acute diarrhoea in 196 cases. Of the 704 wells investigated (sources of drinking-water for the infants), 17.7% were found to be microbially polluted only, 6.3% contained concentrations of nitrate between 50 and 1000 mg/l but were microbially acceptable, and 66.0% were contaminated both microbially and by nitrate.

These findings illustrate clearly that bottle-fed infants receiving feed made up with drinking-water are at greater risk than breastfed infants. Shallow wells are especially at risk from nitrate contamination.

Source: WHO Regional Office for Europe *(102)*.

have been exposed to nitrate above the national standard of 50 mg/l. Data from Ireland also illustrate the increased nitrate levels in small rural supplies compared with large public supplies.

Fig. 5.10. Percentages of samples exceeding the nitrate standard in public and private supplies in Lithuania

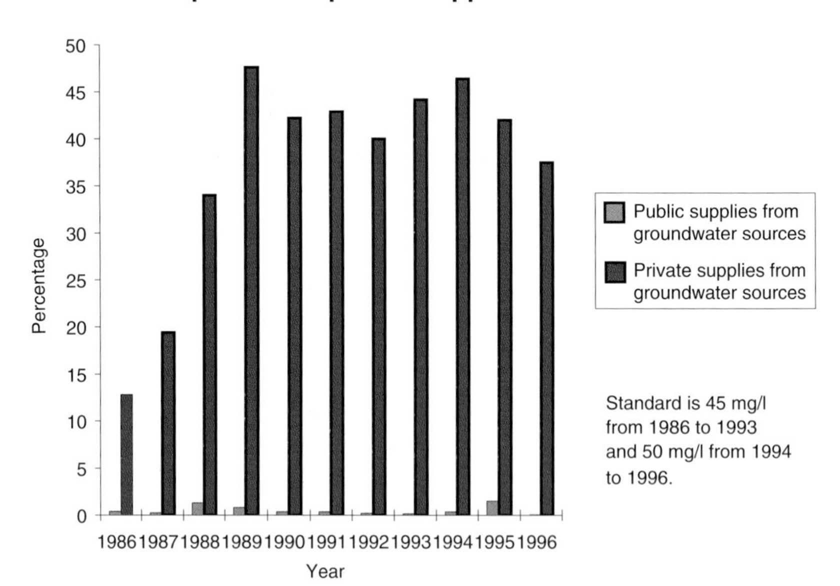

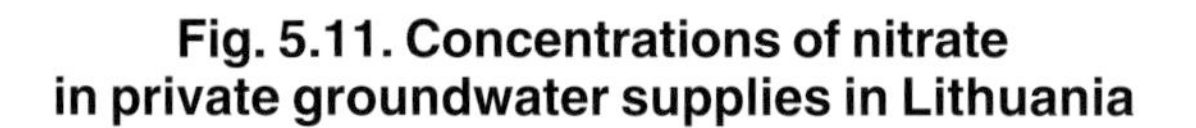

Fig. 5.11. Concentrations of nitrate in private groundwater supplies in Lithuania

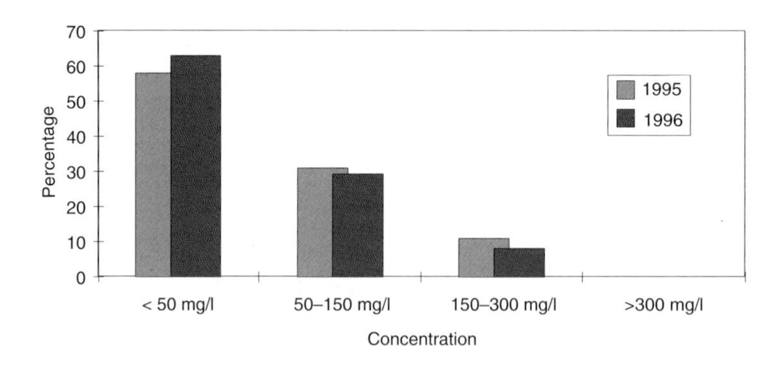

In contrast, Iceland has a stringent standard for nitrate in drinking-water of 25 mg/l, and most samples tested contain less than 1 mg/l.

In Poland, an even stricter standard of 10 mg/l is set for nitrate in drinking-water. This is reported to be exceeded in some cases, primarily in well water in rural areas. Table 5.6 shows the results from

Table 5.6. Percentage of wells exceeding the national standard for nitrates and nitrites in three towns in Cracow *voivodship*, Poland, 1988–1993

Town	Percentage of wells exceeding the national standards	
	Nitrates (> 10 mg/l)	Nitrites (> 0.1 mg/l)
Klaj	39	5
Drwina	56	10
Niepolomice	19	3

the monitoring of 399 samples of well water from three communities located in the Cracow *voivodship* in 1988–1993. In a group of eight infants with acute toxic methaemoglobinaemia treated in Cracow between 1980 and 1994, six had been exposed to nitrates in well water and two to nitrates in carrot soup.

Surface water can also be polluted by nitrate from agricultural runoff and sewage discharges. In a number of countries, a significant percentage of the drinking-water samples tested contained concentrations higher than permitted under drinking-water legislation (usually 45–50 mg/l). In Latvia, nitrate levels rarely exceed 20 mg/l, although at sites of intensive agrochemical use, 55% of boreholes are reported to show an increase compared with natural levels.

Radioactivity

Radioactivity is ionizing radiation released in the form of energy or particles from a material, and chemicals that exhibit this effect are termed radionuclides. When ionizing radiation passes through living tissue, it can interact with the genetic material of the cell, causing mutations that may ultimately lead to cancer. The rate at which the radiation is released from a radionuclide is termed the activity (measured in becquerels, Bq) and the manner in which it is released determines the type of radioactivity: alpha, beta or gamma. The different forms of radiation cause different degrees of biological damage, alpha and beta being most damaging, and different types of tissue vary in susceptibility to damage from radiation.

The contribution of drinking-water to the annual average exposure to natural radiation is very small. Most is naturally occurring radionuclides in the uranium and thorium decay series, including radon. Human activities may increase natural radioactivity levels in water locally, but this is usually from regulated discharges.

Naturally occurring uranium isotopes (uranium-235 and uranium-238) have low radioactivity. Uranium causes inflammation of the kidney at high concentrations, and this chemical toxicity occurs at levels lower than those at which there is a significant carcinogenic risk from uranium radiation. Thus, the radioactivity of uranium is of secondary importance, as a risk to human health, to its chemical toxicity. The provisional WHO guideline level for uranium in drinking-water is 0.002 mg/l *(46)*.

To take account of the spectrum of radionuclides that may be present, the varying biological effectiveness of each type of radiation and the widely varying activity levels of radionuclides, WHO recommends guideline activity concentrations of 0.1 Bq/l for gross alpha activity and 1.0 Bq/l for gross beta activity to screen water for radioactivity *(105)*. If these levels are exceeded, more detailed analysis should be undertaken to determine which radionuclides are responsible and appropriately qualified personnel should assess the risk. Nevertheless, levels of radioactivity much higher than this would still be considered safe for short-term consumption and are preferable to cessation of supply. The method of detection used for gross alpha and beta emitters will not detect volatile radionuclides such as radon, and standard measurement techniques will not detect low-energy beta emitters such as tritium.

Radon-222 is produced as part of the uranium and thorium decay series; where these elements occur in the soil, radon is naturally released from the ground. Radon is an insoluble gas readily released from surface water, which usually has low radon concentrations. Groundwater can contain high levels of radon, especially small private supplies, which may utilize small aquifers unsuitable for public supply and very high in natural radionuclides. Groundwater often undergoes less treatment than surface water, so there is much less opportunity for degassing to occur before the water reaches the tap. The risk from drinking radon in water is considered low, as radon gas and its decay progeny are very short-lived. However, radon has been implicated as a carcinogen when inhaled on a continual basis. Theoretical calculations from Sweden (population about 9 million) suggest that radon in water may cause 35–75 cases of cancer per year. Between 10 and 20 of these are attributed to drinking the water and the remainder to breathing radon gas emitted by water *(103)*. In July 1997, Sweden set new standards for radon in drinking-water. These impose a recommended upper limit of 100 Bq/l and classify water containing more than 1000 Bq/l as unfit for human consumption. As many as 46% of drinking-water samples tested in Sweden contained levels of radon-derived radioactivity above 100 Bq/l, and 3% were higher than 1000 Bq/l.

The contribution of radon from drinking-water to radon concentrations within a house is difficult to determine because of numerous confounding factors such as building construction, the level of ventilation and local geology. One study *(158)* estimated that only 1% of lung cancer in Finland could be attributed to indoor radon, despite a very high mean radon concentration compared with the rest of the world. Exposure to radon via drinking-water may, however, be significant in some local areas, and WHO recommends that total exposure via inhalation and ingestion be taken into account in each situation *(105)*.

Pesticides

Analysis for pesticides is expensive, and the extent of analysis is likely to vary widely, including the number of samples taken and the type of pesticide looked for, both within and between countries (Tables 5.7 and 5.8). No method exists to analyse the “total amount of pesticides”.

The EU countries are bound by the Drinking Water Directive *(47,48)*. The Directive sets limits for individual pesticides at 0.1 μg/l and for total pesticides at 0.5 μg/l, and the revised Directive from 1998 *(48)* introduces more stringent ones for aldrin, dieldrin, heptachlor and heptachlor epoxide of 0.03 μg/l.

Triazines are among the pesticides detected most frequently at levels exceeding the national standards in Belgium, England & Wales, France, Germany, Romania and Sweden. In the United Kingdom,

Table 5.7. Countries in the European Region reporting no problems in meeting national pesticide standards

Country	Comments
Croatia	Concentrations of pesticides are reported to be generally below the gas chromatography detection limit (0.001–0.002 µg/l) and do not exceed the national standard.
Estonia	None of the drinking-water samples tested exceeded the national standard, and only trace levels of pesticides were detected.
Finland	Concentrations of pesticides analysed in drinking-water samples are very low.
Greece	No cases reported in which the national standard was exceeded for total pesticides in the limited number of samples analysed (41 in 1995).
Iceland	Routine monitoring is not carried out. However, analyses have been carried out at the larger waterworks, and pesticides have never been detected in Icelandic drinking-water.
Monaco	No cases reported in which the national standard was exceeded.
Scotland[a]	In 1996, five minor failures were reported relating to pollution incidents, causing three supplies to exceed the limit for individual pesticides.
Spain	None of the drinking-water samples from public supplies tested in 1995 exceeded the national standard of 0.5 µg/l.
Sweden	National statistics are not available, but surveys over the last decade have indicated that levels of pesticides are low even in areas where pesticides are heavily used and with unfavourable water sources. Triazines are detected most frequently.

[a] Scottish Executive *(159)*.

1.6% of drinking-water samples tested for atrazine in 1995 exceeded the national standard of 0.1 µg/l. In France in 1995, 17% of the 2401 samples tested exceeded 0.1 µg/l of atrazine, and 2.1% of the 2756 samples tested for simazine exceeded the standard. More than 16 500 people in Germany are supplied with drinking-water containing atrazine at a concentration above the national standard; levels of up to 3 µg/l have been reported, thus exceeding the WHO guideline value of 2 µg/l. Wells in Romania had triazine concentrations between 0.016 µg/l and 24.41 µg/l, with a mean of 3.66 µg/l.

Table 5.8. Countries in Europe considering contamination by pesticides to be among their major problems in drinking-water quality

Country	Comments
Austria	None.
Belgium (Wallonia)	Triazines and substituted ureas are the main problems. The percentage of samples exceeding the national pesticide standard was 13.4% in 1993 (651 samples), 7.3% in 1994 (1224 samples) and 4.3% in 1995 (1359 samples). However, these are not random samples and therefore have limited statistical importance.
England and Wales	Of the drinking-water samples tested in 1995, 3.2% of samples exceeded the standard of 0.5 μg/l for total pesticides. This figure dropped to 0.3% in 1996. Corresponding figures for the standard for individual pesticides are 0.8% (1995) and 0.2% (1996). Triazines frequently detected; 1.46% of samples failed to comply with national pesticide standard.
France	Atrazine, simazine, desethylatrazine and diuron are the major problems.
Germany	Triazines frequently detected.
Italy	Pollution of drinking-water by herbicides is regarded as an important problem *(91)*; 0.13% of samples failed to comply with national pesticide standard.
Romania	Triazines frequently detected; organochlorines also of concern.
Slovenia	None.

In England and Wales, the uron and chloroalkanoic acid herbicides have also been detected in drinking-water samples: 12.3% of drinking-water samples tested in 1995 exceeded the national standard for isoproturon of 0.1 μg/l. Nevertheless, the significance to human health of concentrations exceeding the guidelines is unclear. The WHO guideline value for isoproturon is 9 μg/l *(105)*. A high proportion (12.8%) of the samples tested in the United Kingdom for 1,1,1-trichloroethane and 4.7% of those tested for dalapon also failed to comply with the national standard. 1,1,1-Trichloroethane and dalapon can both be by-products of chlorination. In Northern Ireland, 4-chloro-2-methylphenooxyacetic acid, mecoprop, isoproturon, simazine and atrazine are the pesticides found most frequently at levels exceeding the national standard.

A number of studies have been carried out on the presence of organochlorine insecticides (α-, β- and γ-hexachlorocyclohexane (HCH), aldrin, dieldrin, heptachlor, DDT and DDE) in drinking-water in Romania. Only 27% of samples from 80 towns in southern Romania contained less than 0.1 μg/l. Concentrations of 0.1–1 μg/l were reported in 6% of the samples, and 37% contained between 1 μg/l and 5 μg/l, 17% 5–10 μg/l and 13% over 10 μg/l. All the 16 samples from Bucharest exceeded the national standard, with a range of 0.54–1.95 μg/l and a mean of 0.92 μg/l. Well water is also contaminated: 64% of samples contained organochlorines at levels above the national standard, with a mean of 0.74 μg/l (range 0.001–4.81 μg/l). For comparison, the WHO guidelines (and the EU limits from 1998) for aldrin and dieldrin are 0.03 μg/l, and the WHO guideline for DDT is 2 μg/l *(105)*.

The Netherlands reports bentazone, atrazine, bromacil, aminomethylphosphonic acid (a metabolite of glyphosate), 4-chloro-2-methylphenooxyacetic acid and diuron as the pesticides most frequently detected at levels exceeding the EU standards under the old Drinking Water Directive.

In Flanders (Belgium), atrazine, simazine, diuron and isoproturon are reported as occurring frequently in raw surface water. Although pesticide pollution is regarded as a problem in achieving high-quality drinking-water, however, pesticides are not reported to be found in drinking-water because it is treated with activated carbon.

Disinfection by-products

The by-products of chlorination have been well investigated and include trihalomethanes and chloroacetic acids. Although several disinfection by-products have been shown to be carcinogenic in animal studies, the overriding priority in providing clean drinking-water must be good microbial quality, to prevent waterborne diseases. Disinfection is the most important step in the treatment of water for public supply and microbial quality should not be compromised in attempting to minimize the formation of disinfection by-products. The level of disinfection by-products can be reduced, however, by optimizing the treatment process. Removing organic substances prior to disinfection reduces the formation of potentially harmful by-products.

WHO guideline values are set for carcinogenic disinfectant by-products. Of greatest importance are those for chloramines (3 mg/l) and chlorine when used as a disinfectant (5 mg/l), bromoform (100 μg/l), dibromochloro-methane (100 μg/l), bromodichloromethane (60 μg/l) and chloroform (200 μg/l). Provisional guideline values have been set for chloral hydrate (10 μg/l), chlorite (100 μg/l), bromate (25 μg/l), dichloroacetic acid (50 μg/l) and trichloroacetic acid (100 μg/l) *(101)*.

The revised EU Drinking Water Directive *(48)* now contains specific standards for disinfection by-products, including 100 μg/l for total trihalomethanes and 0.10 μg/l for epichlorohydrin (see Box 5.5).

In 1995, 3.6% of samples in the United Kingdom contained total trihalomethanes at concentrations above the national standard of 100 μg/l. In the Wallonia region of Belgium, the frequency of concentrations exceeding the national standard was 3.5% (41 in 1171 samples) in 1993 and 9.8% (212 in 2169 samples) in 1995. In Italy, the standard for organohalogenated compounds is 30 μg/l, but an estimated 5% of the population is supplied with water to which a waiver allowing concentrations of up to 50 μg/l is applied *(91)*.

Solvents

Many countries do not have standards for solvents and monitoring is, therefore, unlikely to be carried out regularly. Several countries (Croatia, England and Wales, Hungary and Slovakia) indicate that chlorinated solvents such as trichloroethene and tetrachloroethene may be occasionally detected in drinking-water. However, the concentrations rarely, if ever, exceed the national standards substantially or reach levels considered to be a health concern. In the United Kingdom, only 0.06% of samples tested for tetrachloroethene in 1995 exceeded the national (and EU) standard of 10 μg/l, and no concentrations of trichloroethene exceeding the national standard were recorded. In Hungary, some data indicate the presence of trichloroethene and tetrachloroethene, but at levels that are not considered a health hazard. Dichloroethane is present at the waterworks

Box 5.5. The revised EU Drinking Water Directive

As of 25 December 2003, Directive 80/778/EEC is replaced by Council Directive 98/83/EC on the quality of water intended for human consumption. The objective of the Directive is to protect human health from the adverse effects resulting from the contamination of water intended for human consumption by ensuring that it is wholesome and clean. It has been many years since the original Directive was adopted, however, and much scientific and technical progress has been made since. Under the revised Directive, every five years the Commission shall review the scientific and technical progress made and make proposals for amendments to the Directive.

The Directive aims to improve criteria for assessing and monitoring pollution of drinking-water and to speed up the harmonization of such criteria at European level. The revision was considered necessary not only to update the provisions of the Directive but also to adopt a more integrated preventive approach. The revised Directive focuses on compliance with essential health standards – the scientific benchmark adopted by the Commission was constituted by the 1993 WHO Recommendations on the Quality of Drinking Water – while allowing member states to add secondary parameters if they wish. The old Directive contains 67 parameters, and notable changes in the revised Directive are shown below.

Parameter	Former Directive *(46)*	Revised Directive *(47)*
Boron	300 μg/l	1.0 mg/l
Bromodichloromethane	15 μg/l	Total trihalomethanes 80 μg/l
Chloroform	40 μg/l	
Total pesticides		0.5 μg/l
Endocrine disruptors		New parameter – value not yet determined
Lead	50 μg/l	10 μg/l

supplying one town but at levels below the WHO guideline value of 30 μg/l *(105)*. In the United Kingdom the standard for tetrachloromethane is 3 μg/l, but only 0.01% of samples exceeded this value in 1995.

Aluminium

Aluminium has been implicated in causing three neurodegenerative diseases – amyotrophic lateral sclerosis, parkinsonism dementia and Alzheimer disease. There is support for a positive but not conclusive relationship between the concentrations of aluminium in drinking-water and the incidence of Alzheimer disease *(160)*. No health-based guideline is derived, but WHO recommends a maximum concentration of aluminium in drinking-water of 0.2 mg/l *(105)*.

Aesthetic aspects

Iron and manganese can be found in groundwater, depending on the local geology. In addition, iron salts may be used as coagulants in drinking-water treatment. The ions of both metals can be removed by correctly adjusting the pH in the coagulation procedure. Nevertheless, this stage of treatment is often not applied to groundwater that may contain naturally high levels. Where iron salts are used in coagulation, excessive dosing and poor control of the pH may be the cause of high concentrations in the finished water. The primary concern with excessive levels of these metals in water supplies is their detrimental effect on the aesthetic qualities of the water. Both cause "dirty water" problems, forming insoluble precipitates when they are oxidized.

Owing to the limitations of the animal data as a model for humans, and the uncertainty surrounding the human data, a health-based guideline value for aluminium cannot be derived at this time. The beneficial effects of the use of aluminium as a coagulant in water treatment are recognized. Taking this into account, and considering the health concerns about aluminium (i.e. its potential neurotoxicity), a practicable level is derived based on optimization of the coagulation process in drinking-water plants using aluminium-based coagulants, to minimize aluminium levels in finished water. Thus 0.2 mg/l or less is set as a practicable level for aluminium in finished water *(161)*.

Drinking-water in some areas of Belgium, Finland, Latvia, Lithuania, the Netherlands and Northern Ireland is reported to contain naturally high concentrations of iron and manganese because of the local geology. In Latvia, over half the groundwater intakes are not equipped with filters to remove excess iron, and thus up to 50% of tested water samples do not meet the organoleptic quality requirements. In Hungary, Lithuania, the Republic of Moldova and Sweden, poor water treatment and the unavailability of appropriate treatment technology contribute to elevated levels of iron and manganese in drinking-water. High levels of aluminium and iron in drinking-water in France are attributed to malfunctioning treatment plants. In parts of

Belgium and England and Wales, high levels generally result from iron dissolving in the distribution system. Corrosion of metal piping is also a problem in distribution systems in Norway. In the United Kingdom, 2.4% of samples tested in 1995 exceeded the national standard for iron of 0.2 mg/l, as compared with 5.3% in Ireland. Between 1994 and 1997, about 3% of samples tested in Slovakia exceeded the national standard for iron of 0.3 mg/l. Manganese levels above the national standard of 0.05 mg/l were found in 4.2% of the drinking-water samples tested in Ireland in 1995.

High concentrations of other ions, such as sulfate and chloride, can also impart an unpleasant taste to drinking-water. The total dissolved solids or conductivity is commonly measured in assessing the aesthetic quality of water supplies. Although a high level of dissolved solids does not inherently pose a health hazard, the risk occurs if water intake is substantially decreased because it is not as palatable. Similarly, excessive quantities of iron and manganese, which also affect the aesthetic quality of the water, might be expected to have similar effects.

Flooding, drought and disease

Floods

Much of the effect of flooding on mortality and ill health may be attributed to the distress and the mental effects of the event. Following flooding in Bristol, United Kingdom, a 50% increase in mortality was reported among those whose homes had been flooded. In addition, primary care attendance rose by 53% and referrals and admissions to hospital more than doubled *(162)*. Similar mental effects were found following floods in Brisbane in 1974 *(163)*. Mental symptoms and post-traumatic stress disorder increased, including 50 flood-linked suicides, in the two months following major floods in Poland in 1997 *(164)*.

During and following both catastrophic and non-catastrophic flooding, health is at risk if the floodwater becomes contaminated with human or animal waste. Floods in Europe are associated with an increased risk of leptospirosis. An epidemic of leptospirosis was associated with floods in the Ukraine in 1997. Flooding in Lisbon in

1967 was associated with a small outbreak of leptospirosis *(165)*. Analysis of surveillance data following the major floods in the Czech Republic in 1997 suggests an increase in cases of leptospirosis *(166)*.

Several studies have established a link between dampness in the home, including occasional flooding, and a variety of respiratory symptoms. For example, flooding has been significantly linked to childhood experience of cough, wheeze, asthma, bronchitis, chest illness, upper respiratory symptoms, eye irritation and non-respiratory symptoms *(167)*. Very little is known about the occurrence of other diseases, such as skin diseases.

As a general rule in Europe, the floods that bring about the highest material losses are different from those that claim the highest number of deaths. Floods responsible for a high number of deaths largely take place in open spaces in mountainous basins, where there is little time to respond (in many cases less than 3 hours) and where the surprise factor is the main cause of death. However, these phenomena rarely cause major material losses. Floods in urban areas do not generally occur suddenly, and there is usually capacity for real-time forecasting, so that the population can be alerted and evacuated in advance.

Mainland Spain has experienced an average of five floods per year over the last five centuries and the average cost of material damage per year is estimated to be US $6 billion *(168)*. Recent devastating floods in central Europe affecting the Czech Republic, Germany and Poland killed over 400 people and left many thousands destitute. Some 160 000 were evacuated from their homes in Poland and a further 50 000 in the Czech Republic *(164)*.

Drought

No deaths have been caused directly by a shortage of water in Europe over the last 50 years. Nevertheless, severe restrictions may result in a lack of hygiene and the use of resources that are not subject to sanitation control. Not only is there a lack of statistics on this type of effect, but the quality and quantity of the water are often not considered holistically. The absence of statistics is in part the result of the lack of a uniform definition of drought and, unlike floods, evaluating or quantifying a drought is difficult. Criteria are needed that would clearly distinguish between and define drought, desertification, water shortage and water scarcity.

Recreational water, health and disease

Recreational water environments have a diverse range of hazards to human health. These include factors associated with accidents (resulting in drowning and near drowning and spinal injuries), microbial pollution, exposure to toxic algal products and occasional exposure to chemical pollution *(159)*. Hazards exist even in unpolluted environments – eye infections in bathers probably occur as a result of a reduction in the eye's natural defences through contact with water *(169)*.

Drowning and physical injuries

Information on accidents is not systematically collected in all countries throughout Europe. Deaths from drowning are available from the WHO mortality database, but the accidents related to bathing water are not separately recorded in most cases (Table 5.9) *(105)*. Data suggest that males are more likely to drown than females, but it is not clear whether this is because more males swim (Fig. 5.12) *(157)*.

The greater consumption of alcohol by men is one contributing factor *(171)*. However, heart attack, sea currents and surf also contribute to drowning accidents. Private pools, lakes and other bodies of fresh water contribute significantly to drowning statistics, especially in children. In Denmark, for example, 63% of all drownings among 0- to 14-year-olds between 1989 and 1993 occurred in these types of water body. In terms of all accidental deaths in the European Region, drowning accounts for less than 10% of the 280 000 deaths due to accidents *(170)*.

Permanent paralysis may occur as a result of diving into a shallow body of water or swimming pool. A spinal cord injury can result in paraplegia or quadriplegia, the latter being more common. The number of spinal injuries sustained as a result of swimming accidents does not appear to be widely available. Data available from the Czech Republic suggest that spinal injuries are more frequently

Table 5.9. Mortality rates from accidental drowning and submersion per 10 000 population in 38 countries in the European Region in 1994 (unless stated otherwise)

Country	Males	Females
Albania (1993)	0.49	0.18
Armenia (1992)	0.24	0.05
Austria	0.27	0.08
Belarus (1993)	1.34	0.24
Belgium (1992)	0.15	0.06
Bulgaria	0.56	0.07
Croatia	0.47	0.08
Czech Republic (1993)	0.44	0.13
Denmark (1993)	0.12	0.03
Estonia	2.87	0.49
Finland	0.54	0.06
France	0.17	0.04
Germany	0.10	0.05
Greece	0.41	0.12
Hungary	0.55	0.09
Ireland	0.42	0.04
Israel	0.14	0.06
Italy (1993)	0.16	0.03
Kazakhstan	1.30	0.29
Kyrgyzstan	1.44	0.39
Latvia	3.77	0.71
Lithuania	3.73	0.55
Netherlands	0.11	0.03
Norway	0.29	0.06
Poland	0.78	0.15
Portugal	0.12	0.04
Republic of Moldova	1.79	0.38
Romania	1.16	0.29
Russian Federation	2.13	0.34
Slovenia	0.43	0.12
Spain	0.29	0.06
Sweden	0.21	0.03
Switzerland	0.18	0.08
Tajikistan (1992)	0.57	0.24
Turkmenistan	1.12	0.39
Ukraine (1992)	1.62	0.27
United Kingdom	0.08	0.02
Uzbekistan (1993)	0.73	0.34

Source: World Health Organization *(105)*.

sustained in open freshwater bathing areas than in supervised swimming areas. However, the number of injuries sustained in freshwater areas in the Czech Republic declined from 38 in 1995 to 30 in 1997.

Fig. 5.12. Numbers of people drowning in Israel and Italy by sex, 1979–1987

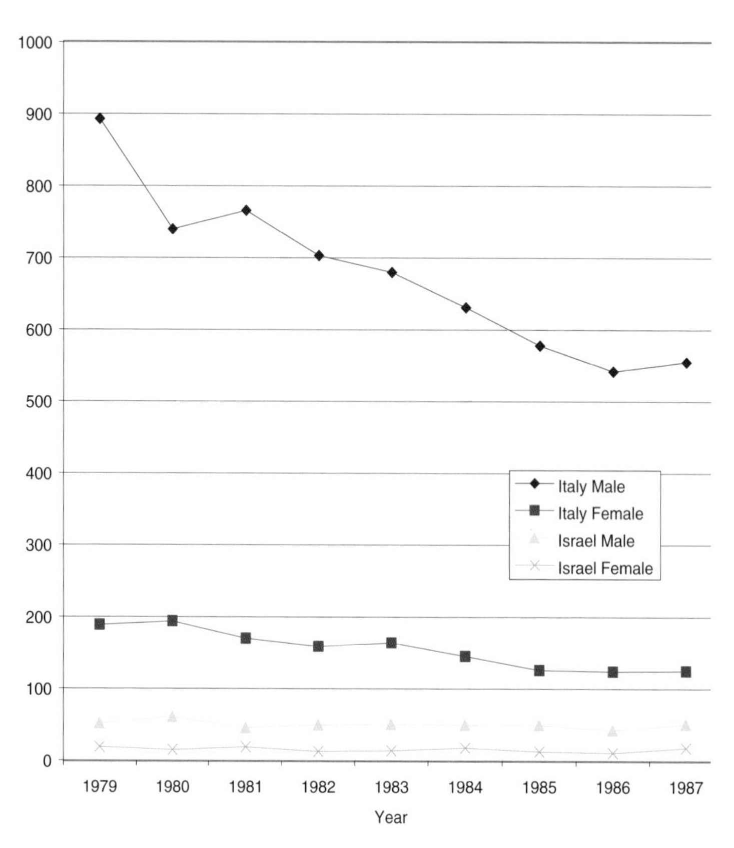

Source: World Health Organization *(170)*.

Microbial factors

Recreational water may contain a mixture of pathogenic and non-pathogenic microbes from a variety of sources, including human sewage effluent, industrial processes, farming activities, wildlife and recreational water users. In addition to these sources, other microbial and biological hazards such as leptospires and algal blooms may present a risk. Clear evidence indicates that exposure to faecal pollution through contaminated recreational water leads to detectable health effects *(157,172)*. Microbial contamination of bathing water, primarily in the Mediterranean Sea, is responsible for more than 2 million cases of gastrointestinal disease annually *(6)*.

Several studies have provided evidence of a dose–response relationship between enteric and nonenteric illnesses and faecal pollution *(172–175)*. Enteric illness, such as self-limiting gastroenteritis, is the most frequently reported adverse health outcome investigated, and evidence suggests a causal relationship between increasing recreational exposure to faecal contamination and frequency of gastroenteritis *(157)*. Associations between ear infections and microbial indicators of faecal pollution and bather load have been reported, and increased rates of eye symptoms have been reported among bathers. Evidence also exists for more severe health outcomes. Although the probability of contracting acute febrile respiratory illness is generally lower than that of contracting gastroenteritis, faecal pollution and acute febrile respiratory illness may have a cause-and-effect relationship. There is reason to believe that other severe infectious diseases such as typhoid fever and viral diseases such as hepatitis A and E may be transmitted to susceptible bathers who make recreational use of polluted water *(158)*.

Monitoring of the quality of bathing water in the WHO European Region is based on the definition of the bathing area as passing or failing a defined microbial standard. This approach has severe limitations, as an identified bathing area can vary widely in spatial and temporal factors. Data on the microbial quality of recreational water are collected mainly by regulatory agencies to assess compliance. Problems of analytical reproducibility and interlaboratory comparability, and variations in the number of water bodies monitored, sampling frequencies and temporal conditions make comparison between countries difficult.

An improved approach to the regulation of recreational water is needed that better reflects the health risk and provides enhanced scope for effective management intervention. EU member states monitor designated bathing waters for compliance with the directive on the quality of bathing water *(176)*, and some non-EU countries monitor routinely (Table 5.10). Published water quality data are always retrospective because of the nature and frequency of analysis undertaken. An alternative approach to monitoring and assessment is to classify a beach based on health risk, as opposed to the current pass-or-fail approach in place in the EU. The health risk approach is more flexible than the pass-or-fail approach, and allows

Table 5.10. Number of bathing sites and percentage complying with national microbial standards in selected non-EU countries

Country	Number of recognized bathing sites		Percentage of coastal (a), freshwater (b) and all sites (c) complying with national microbial standards			
	Freshwater	Coastal	1994	1995	1996	1997
Croatia	158	831	(a) 91.9 (b) 95.8 (c) no data	(a) 93.5 (b) 96.5 (c) no data	(a) 93.6 (b) 97.7 (c) no data	(a) 96.7 (b) 96.8 (c) no data
Czech Republic	276 (open areas) 237 (public pools) 549 (seasonal)	None	(a) no data (b) no data (c) 84–86			
Finland	360	94	(a) no data (b) no data (c) no data	(a) no data (b) no data (c) 98.2	(a) no data (b) no data (c) 99.7	(a) no data (b) no data (c) 98.5
Lithuania	149	10	(a) no data (b) no data (c) 55.9	(a) no data (b) no data (c) 58.5	(a) no data (b) no data (c) 70	(a) no data (b) no data (c) 61.3
Republic of Moldova	42	0	(a) no data (b) 63.7 (c) no data	(a) no data (b) 41.1 (c) no data	(a) no data (b) 63.8 (c) no data	(a) no data (b) 60 (c) no data

managers to respond to sporadic events and reclassify a beach. It provides a more robust, accurate and feasible index of health risk, by combining a measure of faecal contamination with an inspection-based assessment of the susceptibility of an area to direct influence from human faecal contamination *(177)*. A beach can receive higher classification if human exposure is reduced at times or places of increased risk *(177,178)*.

More coastal bathing sites than freshwater sites are designated within the EU, despite the common use of freshwater sites for recreational activities (Fig. 5.13). The quality of freshwater sites designated for bathing is considerably worse than that of coastal sites in the EU, although the overall quality trend for both coastal and freshwater sites appears to be improving.

Fig. 5.13. Numbers of seawater and freshwater points in the EU sampled for compliance with the directive on the quality of bathing water, 1991–1997

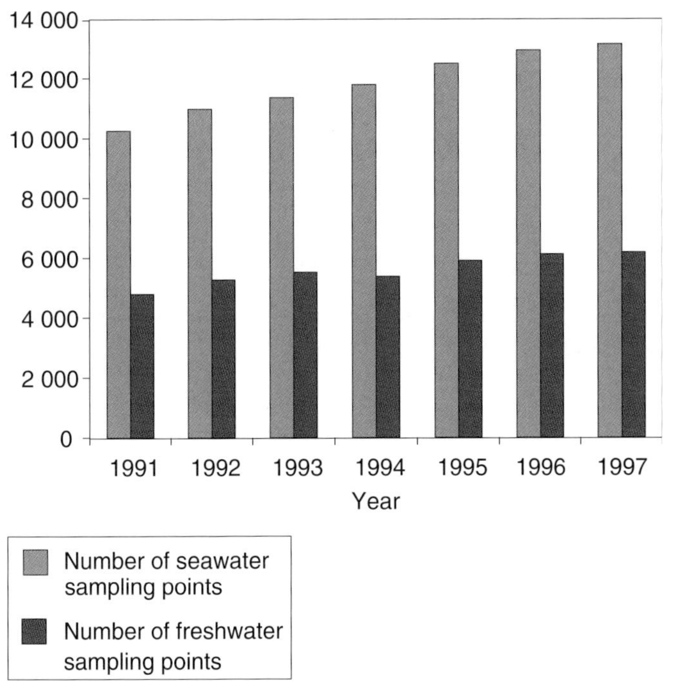

Source: European Commission *(179)*.

Data on other parameters related to recreational water quality across Europe are available sporadically. Physicochemical, aesthetic and other biological parameters are not routinely monitored, and international comparability is thus not possible.

Cyanobacteria

Coastal marine environments have many toxic species of cyanobacteria, and several types of human health effect have been reported in association with cyanobacteria *(36)*. The evidence documenting the health effects for recreational water users from marine toxic phytoplanktonic blooms originates primarily from the Black Sea, the Baltic Sea, the North Sea and parts of the Mediterranean and Adriatic Seas. Even minor contact with cyanobacteria in bathing water can lead to skin irritation and an increased risk of gastrointestinal symptoms *(180)*. Observations of human deaths through cyanobacterial toxins have been limited to exposure through renal dialysis *(157)*, but there are many reports of animal deaths from cyanobacterial toxicity in North and South America, Europe, Australia and Africa *(181,182)*. The human health implications are primarily known from anecdotal reports of irritations of the skin and/or mucous membranes. Marine cyanobacterial dermatitis may occur after swimming in seas containing blooms of certain species of marine cyanobacteria *(183)*. Inhalation of sea spray aerosol containing fragments of marine dinoflagellate cells and/or toxins released into the surf by lysed algae can be harmful *(184,185)*. As the evidence of risks to human health associated with the presence of cyanobacteria during recreational activities is limited, there are currently no guideline values *(157)*.

Shellfish

The main types of human exposure to pathogenic microorganisms in the marine environment are through direct contact with polluted seawater, sand or sediment and through consumption of contaminated shellfish. The potential health risks arising from consumption of shellfish relate to the quality of the water in which the shellfish are grown or harvested and the quality of the shellfish when they reach the market. The link between shellfish water quality and the health effects on humans as the final consumers is more complex than that applying to bathing water. The extent of faecal contamination that can be tolerated in the growing water is complex, and there is no satisfactory correlation either between bacterial indicator levels in shellfish and those in the growing water, or between the indicators and pathogens in the shellfish themselves *(186)*.

Overall morbidity statistics are insufficient, as practically all diseases caused by pathogens are capable of being contracted through media other than the marine environment. For example, in 1974, a large outbreak of cholera affected several areas of Portugal. There were 2467 bacteriologically confirmed cases leading to hospital admission, and 48 deaths. The most significant risk factor for the national outbreak was the consumption of raw or semi-cooked shellfish, although epidemiological investigations in Lisbon showed that visitors to the local spa had a much higher attack rate *(187)*.

Artificial purification of mussels is widely practised throughout Europe. One method involves the use of chlorine to disinfect the seawater, which must then be dechlorinated before it can be used to depurate contaminated shellfish. Although this method is relatively expensive, it is used widely in Spain and France. Disinfection by ozone is now the depuration method of choice in large shellfish-cleaning stations in France, but a considerable amount of shellfish throughout Europe is still not subject to strict depuration procedures or proper control of storage after harvest *(188)*.

VECTOR-BORNE DISEASES

Malaria

Malaria is by far the most important water-related, vector-borne disease, in terms of both numbers of sufferers and of directly attributable deaths. The causative agents in humans are four species of *Plasmodium* protozoa. Of these, *P. falciparum* accounts for most infections and is the most lethal.

The geographical area affected by malaria shrank considerably during the second half of the twentieth century, but control is becoming more difficult. Increased risk of the disease is associated with changes in land use linked to activities such as road-building, mining, logging and agricultural and irrigation projects. Other causes of its spread include global climate change, disintegration of health services, armed conflicts and mass movements of refugees. Increasing travel is causing malaria to re-emerge in areas from which it had previously been eradicated, such as Azerbaijan and Tajikistan. In the United Kingdom, 2364 cases of malaria were registered in 1997, all imported by travellers. Malaria is currently epidemic in Armenia, Azerbaijan, Tajikistan, Turkey and Turkmenistan.

6

Present policies

Both the quantity and quality of water need to be considered in achieving sustainable water management. A balance has to be achieved between the abstractive uses of water, in-stream uses such as recreation and ecosystem maintenance, the discharge of effluents and the impact of diffuse sources of pollution. Surface water systems are now often highly regulated in efforts to control water availability, whether for direct use in irrigation, for hydroelectric generation or drinking-water, or for protecting against the consequences of floods or droughts. Changes in the nature and scale of human activities influence both the qualitative and quantitative properties of water resources. The change in trends in water use from rural areas and agriculture to urban areas and industry is reflected in patterns of demand and water pollution. It is unclear whether this shift will continue in the future; irrigated agriculture demands increasing amounts of water, a process that already accounts for around 70% of water demand worldwide. Many industries have successfully developed processes with substantial water economy measures. Development of appliances that use less water and control of losses from water mains may therefore stabilize or even reduce total demand for water in the future *(37)*.

Quantity management

Water consumption has been high in recent decades, stemming from the perception that water is abundant and inexpensive. This perception is changing in the light of increasing water pollution, drought or water stress and rising water prices. In many countries, the enforcement of EU or national legislation and the rising cost of water are leading municipalities and industries to reduce water use and to invest in water-saving processes and equipment.

The two main approaches to increasing the efficiency of the supply and use of water are demand management, which includes economic incentives to reduce use and increase the efficiency of use, and infrastructure changes. These changes may take the form of improving the network, increasing storage capacity, improving industrial efficiency, using low-consumption devices and new sources, recycling and water transfer.

Use of available resources

The EU's fifth programme of policy and action in relation to the environment and sustainable development of 1993 *(189)* encourages member states to employ environmental taxes and charges to achieve a more cost-effective environmental policy. This has led to an increase in the application of a wide range of different charging or taxation policies in the member states, including charges for water abstraction and use.

Abstraction charges

A system of licensing has traditionally been applied to try to achieve the required balance between the various demands on the aquatic environment: the command and control approach. Nevertheless, as water resources of adequate quality become more scarce and the public interest in the aquatic environment (in terms of minimum flow, quality and aesthetic appearance) increases, economic instruments are increasingly being applied to complement the licensing system.

The charges made for abstracting water must be set at a rate that reduces abstraction without causing undesirable economic or social effects, such as reducing the competitiveness of industry or standards of hygiene in poor households.

The purpose and design of charging schemes for water abstraction vary widely in different countries, reflecting different institutional arrangements and geographical conditions (Table 6.1) *(190)*. The use of monies raised also varies in the different countries, from providing funds for water resource infrastructure to subsidizing water-supply systems. Table 6.2 shows the range of charges applied in selected EU countries.

Table 6.1. Purpose of water abstraction charges in selected European countries

Country	Recover costs	Raise revenue	Provide incentive	Replace taxation
France		Yes	Yes	
Germany	Yes		Yes[a]	
Netherlands		Yes[b]	(Yes)[c]	Yes[d]
United Kingdom	Yes			

[a] Depends on the charging scheme in the individual *Länder*.
[b] Provincial tax for water abstraction on groundwater only.
[c] The charge is for groundwater only.
[d] Introduced in 1995 for general tax-raising purposes: environmental fiscal reform.

Source: Buckland & Zabel *(190)*.

Table 6.2. Range of abstraction charges applied in selected EU countries

Country	Charge (€ per m^3)
France	0.01–0.02
Germany	0.02–0.53
Netherlands	
National	0.15
Provincial	0.08
United Kingdom	0.006–0.021

Source: Krinner et al. *(52)*.

Variation in the charges made for abstraction from different types of water source or for different uses can be an additional instrument for managing demand, and can encourage changes in the pattern of water abstraction (Table 6.3). In Germany, for example, higher charges are generally applied to abstraction from groundwater than from surface water, especially for uses other than potable supply; in some *Länder* charges are made only for groundwater abstraction. Charges may also vary according to the availability of the water, the season of abstraction and how much of the abstracted water is returned to the source, as in the United Kingdom; in France, however, the charges vary according to the quality and vulnerability of the source *(52)*.

Where the abstraction charges are relatively low compared with the price of drinking-water, such as in England and Wales, France, Germany and the Netherlands, the incentive to reduce the volume of water used is low. Current water abstraction charging schemes are best described as financial instruments to raise revenue to cover costs or to fund specified activities, rather than as economic instruments to change the behaviour of water users. None of the charging schemes attempts to set the charges based on the site-specific real value of the water resource.

Although water abstracted for irrigation is increasingly becoming a major issue, charges for such water use tend to be very low – much lower than charges for other uses – and do not represent the real value of the water, especially as irrigation tends to be practised in the summer when water resources are under the greatest threat *(51)*.

Water pricing and metering

Charging can undoubtedly influence the volume of water used. It has been shown that when water charges are raised, consumption decreases. In eastern Germany the consumer price index increased from 1985 to 1992 by approximately 14% for households and 9% for other consumers. Water consumption dropped 10% per annum between 1980 and 1991. Between 1990 and 1995 water consumption per person declined by 9% in Germany as a whole, and the observed drop in water consumption is influenced by changes in consumer behaviour as well as changes in water prices *(52)*. In Croatia, average consumption has fallen from 266 to 150 litres per person per day because of the introduction of higher prices *(50)*. In contrast, water tariffs in Albania are very low, and this has been identified as a factor in

Table 6.3. Water abstraction charges in selected EU countries and regions

Country	Variation in charges	Comments
Germany	Higher charges for groundwater abstraction, especially for uses other than potable supply	
	Hamburg: Charges only applied to groundwater abstraction	Charging scheme resulted in significant refund on unused water rights
	Hessen: Charges only applied to groundwater abstraction	Water consumption reduced by 11% (but this may have resulted, in part, from the influence of the economy)
Netherlands	Charges only applied to groundwater abstraction	Two charges made: • provincial: to finance research for groundwater resource development and water planning • national: part of a reform of the national tax system to shift the burden of taxation from income tax towards consumption
England and Wales	Annual charge based on licensed volume taking into account: • source and region • season • loss factor	Charges levied are usually low and are based on covering the administrative cost of the regulatory authority
France	Charges based on: • volume abstracted • scarcity of water resources • loss factor • charges for groundwater tend to be higher • charges higher for water taken from upper reaches of rivers	The six water agencies collect charges at the catchment level Funds are reallocated to improve water management

Source: Krinner et al. *(52)*.

the increasing demand for water *(93)*. In some cases, rising water prices may lead to the use of alternative water supplies that may be of poorer quality, compromising the health of the population (Box 6.1).

A balance needs to be struck between prices high enough to reduce excessive demand and the health of the population. Supplying water and sanitation services to, for example, small communities in rural areas is difficult and costly, and such communities account for a considerable proportion of the population in some countries *(96)*.

Charges for water generally reflect, to a varying degree, the costs of abstracting, treating and supplying it to the consumer. The costs of constructing resource/source protection are rarely incorporated. Nevertheless, charges can also discourage waste and encourage the use of water-efficient appliances. Several studies have demonstrated that rising water prices positively affect efforts to conserve water, both indoors and outdoors *(191,192)*.

Domestic metering is widespread in many countries (such as Denmark, France, Germany, the Netherlands and Portugal) but less common in others, including the United Kingdom and non-EU countries. The use of metering allows charges to be based on the volume of water actually used, and the charge can therefore be related to the cost. There are incentives to avoid wasting water, and this seems to be borne out in the United Kingdom, where metered households use an estimated 10% less water than do households without a meter.

Box 6.1. Water charges and consumption in Hungary

An analysis of the relationship between water price and water consumption was carried out in three areas in Hungary. Budapest is the largest city with a heavy concentration of industries and vast suburban areas with housing and gardens. Miskolc is the third largest city and also has a heavy concentration of industry, while Fejér County is characterized by high living standards and little industry. From 1987 to 1992, the price of water rose ten-fold in Budapest and more than 20-fold in the other towns. In parallel, water consumption decreased by between 5% and 28% for households and 20–30% for industry, depending on the price before the increase. A secondary, unwanted effect was an increase in the use of private wells, often yeilding water of unacceptable quality. The pattern of decline in water use suggests that a price increase might have a much greater effect on consumption in countries and areas where the price of water had previously been low than in those where water had always been relatively expensive.

Source: Krinner et al. *(52)*.

However, metering increases the capital costs of water supply significantly. Metering requires efficient administrative procedures and a sophisticated maintenance programme to avoid technical problems with the meters. Table 6.4 shows some of the advantages and disadvantages of metering.

Table 6.4. Advantages and disadvantages of metering water

Advantages	Disadvantages
Increase in revenue	Cost
Equity	Irregular income
Reduction in misuse and waste	High levels of underregistration and other technical problems
Conservation of the resource	Logistical difficulties related to inspection and reading
More accurate economic costing and pricing	High level of accuracy required before computerization
Differential tariff structures according to volume consumed	Billing system not adapted to equity objectives
Better technical control of water-supply systems	

Source: WHO Regional Office for Europe *(102)*.

Improving efficiency of use

In most countries the instinctive response to increasing demand for water is to consider increasing the supply. Most authorities respond to water scarcity by non-pricing mechanisms, such as rationing, prohibiting types of use or cutting off supplies. Reforming public behaviour towards water is a difficult task with political and administrative costs. Efficacy is related to the criterion of acceptability, and a combination of measures might be most effective in many cases. Users need to accept responsibility, and all stakeholders must be involved in formulating and implementing policy.

Urban water use

The conservation of water in urban areas can be facilitated in a number of ways *(52)*.

- *Infrastructure programmes*. Losses from the distribution system are reduced by improving the network, repairing weak points or installing measuring devices.

- *Programmes to improve efficiency*. Water consumption is lowered by technically modifying installations, such as by improving sanitary equipment and designing public and private gardens in a way that allows water demand to be reduced.

- *Substitution programmes*. Alternative sources of water are found for uses other than drinking, rather than supplying drinking-water from the public network.

- *Water conservation programmes*. Water consumption is reduced through user education, public awareness, tariff systems and information campaigns.

- *Management programmes*. These comprise municipal regulations, tariff systems, commercial incentives and discounts for economic water use, hydraulic audits, and loans and subsidies for improvement measures.

Various countries have used combinations of these measures (Box 6.2), which contributes to the expectation that water demand in much of the WHO European Region will stabilize or decline.

Agricultural water use

The main use of water for agriculture in the WHO European Region is for irrigation, and this comprises a significant proportion of the total demand in some countries. Some irrigation systems are small-scale, belonging to and managed by an individual farm. The demand for groundwater may be reduced in these cases by constructing individual reservoirs to store winter rainfall, which can then be used for irrigation during the summer. Farmers in south-eastern England are increasingly following this approach.

Large-scale irrigation systems may include storage reservoirs and major canal distribution networks, and their administration often involves public bodies. The efficiency of water use differs greatly among irrigation systems, for example between earth canals and

Box 6.2. Managing urban demand for water in Brittany

The Brittany region of France had difficulty in providing drinking-water for several years because of limited quantities and poor quality, due especially to nitrates and pesticides. A new water supply and management scheme was therefore initiated in 1990. Pilot action plans in several towns led to a reduction in public water consumption of 76% in 15 years (with a stable population) through the installation of low-flow toilets, watering equipment for public gardens, and water-saving appliances in swimming pools and schools. The predicted development of peri-urban zones and the associated increase in water use led the regional and municipal authorities to launch new pilot programmes to reduce domestic water use. In seven towns, the installation of water-saving taps, showers and toilets reduced the annual water consumption by 31 m^3 per household. In addition, the use of new garden watering systems led to a reduction of 60% in water use during the summer. In some towns water consumption was reduced by 50% after 10 months. This pilot action has also increased public awareness in an area where water quality is frequently a major problem.

It is intended that the scheme will be extended beyond the region to make the general public and the plumbing profession more sensitive to water use and water-saving equipment, including testing existing appliances, installing new water-saving equipment in households, identifying and diagnosing losses by waterworks and improving the efficiency of water consumption. Such action programmes have also been conducted in several industries and in agriculture.

Source: Krinner et al. *(52)*.

gravity irrigation and systems employing concrete-lined canals and pipes. There is scope for improvement both in the distribution system and in the efficiency of use at the farm level. Targeting the irrigation where it is most required, levelling fields, improving drainage and choosing the most appropriate irrigation system are local measures that will increase the efficiency of water use *(52)*.

Organizational and management changes such as improving knowledge about water losses, establishing information systems, improving the determination of crop demand, adjusting water allocation, and optimizing timing and tariff systems can all contribute to managing the demand for irrigation water *(52)*.

The potential for using secondary water effluents after adequate treatment might also offer substantial water savings with secondary benefits for water resource quality.

Industrial water use

The mechanisms to reduce demand for water by industry are similar to those that apply to urban water supply, and improving infrastructure and reducing consumption by increasing efficiency and tariffs can thus be beneficial. Recycling and reusing water are important in saving water in industry.

Legal and regulatory frameworks and economic incentives are the measures that are likely to achieve the most success in reducing the industrial demand for water. In addition, programmes aimed at promoting water reuse and recycling can lead to major savings. Rationalizing production in larger units also tends to reduce the consumption of water per product unit *(52)*.

Quality management

Water pollution control

High-quality drinking-water demands that the contamination of source water is avoided. Source protection is often less expensive and more reliable than treatment, and is universally relevant to providing safe drinking-water. In other areas, where advanced treatment is possible, high-quality source water reduces the cost of providing drinking-water.

Many countries in the WHO European Region identify the problem of poor protection of source water from contamination by agricultural activities and industrial or domestic discharges, resulting in microbial or chemical contamination. Both surface water and groundwater are regarded as poorly protected in Sweden; in Hungary, shallow aquifers and bank-filtered supplies are considered particularly vulnerable.

Various measures can be used to prevent microbial and chemical pollution of water resources from both diffuse and point sources, but financial or regulatory incentives may be necessary to ensure their implementation.

Agricultural pollution

Agriculture is carried out over much of Europe's land mass, and agricultural land constitutes an extensive area that is a potentially diffuse source of chemical and microbial pollution. The intensification of agriculture that occurred in the middle decades of the 20th century was encouraged in EU countries by the common agricultural policy, which continues to subsidize production and has substantially increased agricultural production in the EU.

The common agricultural policy was revised in 1992 and production-based subsidies were replaced with direct grants related, for example, to cultivated area. Payments for land to be taken out of production (set aside) were also introduced as an incentive to reduce the intensiveness of agriculture. In addition, a number of other EU environmental schemes related to agriculture are specific incentives to prevent the pollution of water bodies.

The cost of agrochemicals and the economic situation of a country are major determinants of the cost–effectiveness and extent of use of agrochemicals. Taxation or levies on the purchase of pesticides and inorganic fertilizers are another potential management tool. Payments for conversion to organic agriculture are available in some European countries. The economic balance of organic farms is frequently good because of reduced expenditure on agrochemicals and, in some cases, better prices for the product. Cooperation between water supply agencies and farmers in Germany has supported this trend *(37)*.

A change of attitude and community awareness of quality criteria and the ecological effects of products are important in controlling pollution. Governments can use a wide array of measures to encourage this, such as training and advising farmers, conducting "eco-audits" on products, subsidizing the setting aside of land, providing subsidies or tax exemptions during periods of transition to organic farming, imposing pollution taxes, and enacting legislation to enforce water protection *(37)*.

Pesticides

Avoiding the direct spraying of pesticides on to bodies of water is a basic precaution in avoiding contamination. Except for those used to control aquatic weeds or insects, pesticides should not be used in a way that makes contamination of water likely. The conditions of use for some pesticides approved in the EU impose a minimum distance between spraying operations and water. Establishing protection zones around the water sources and regulating practices allowed or prohibited within these zones will reduce the likelihood of contamination. In the Netherlands it is predicted that this method, along with measures to reduce atmospheric deposition and leaching, will reduce water pollution by pesticides by 70–90%.

Reduced pesticide use will result in a lower potential for run-off and leaching. Application rates can sometimes be reduced, without compromising efficacy, by improved application methods such as low-volume spraying and placement applications.

Integrated pest management allows pests to be controlled with fewer chemicals, by integrating a number of pest control techniques. Integrated pest management methods include using crop varieties that are resistant to pests and encouraging populations of beneficial insects that prey on crop pests. Cultivation methods that reduce the need for pesticides, such as the use of reduced sowing rates to reduce the incidence of fungal diseases, are also used.

Fertilizers

Codes of good agricultural practice have been developed in many countries, and most include recommendations designed to reduce the potential for leaching of nutrients from fertilizers. Recommended measures include *(33)*:

- timing fertilizer applications to avoid high-risk times such as autumn, winter and before heavy rainfall;
- applying fertilizers when uptake by the crop is greatest;
- measuring the nutrient balance of the field to allow an informed assessment of the fertilizer required;
- using cover crops during winter months to fix soil nitrogen and cover bare soil;
- using straw fertilization during the winter, as carbon-rich substances in the straw tend to immobilize the soluble soil nitrogen;
- limiting irrigation during high-risk periods;
- using appropriate methods and timing of tillage;
- rotating crops; and
- observing buffer zones alongside watercourses, preferably planted or overgrown, to take up nutrients and reduce surface water contamination.

Animal husbandry

As with agricultural chemicals, the spreading of animal wastes on to land should be timed to minimize the likelihood of run-off and leaching of nutrients and microorganisms. Using less dense stocking regimes ("extensive" farming) alongside watercourses is a measure designed to reduce the contamination of surface water by animal faeces.

Many animal feeds contain concentrations of nutrients that are higher than required by the animals and the excess, rather than being assimilated, is excreted. Optimizing the nutrient content of feeds is an approach that may contribute to successfully reducing the nutrients reaching surface water and causing problems of eutrophication.

Economic instruments

Agriculture

Several countries have introduced tariffs and incentives to encourage the use of farming methods that reduce the potential for contamination of water sources by agricultural chemicals or animal wastes (Boxes 6.3 and 6.4). Prosecution in the case of contamination and incentive schemes to encourage less intensive and/or organic farming are some examples.

Industrial contamination

Regulatory powers and economic incentives are required to reduce the contamination of water resources by industrial activities. Many countries make charges for the discharge of industrial effluent both to sewers and to surface waters and, in EU countries, the discharge of industrial wastewater to surface waters requires a permit from a regulatory body. The permit takes into account the amount and toxicity of the contaminants contained in the wastewater. However, the charges levied may not always reflect these factors, and the effluent charging schemes applied in the different EU countries vary widely (Table 6.5).

Local management

In most countries, local (municipal) authorities are responsible for operating and managing the water supply. In a few countries, notably

Box 6.3. Legal instruments and state subsidies to encourage lower levels of agrochemical use in Switzerland

Agriculture in Switzerland is undergoing considerable change, characterized by a move towards integrated farming that minimizes the use of chemical products. To encourage this trend, legislation was introduced in 1993 to provide for payments to farmers who use ecologically sound farming methods. To qualify for these payments, farmers must comply with strict requirements on reducing agrochemical input to the land by demonstrating compliance with the requirements of integrated production or organic farming.

The requirements of integrated production focus on "extensification" of farming, with minimal use of fertilizers and pesticides, and safeguarding or increasing natural biodiversity on agricultural land. For example, no preventive pesticide application is permitted and, for cereal production, no fungicides, pesticides or growth regulators are allowed, although some use of herbicides is permitted. For both the integrated production and organic farming schemes, codes of practice have been produced and there are strict controls (inspections, spot checks and analyses) to verify that farmers qualify for payments. In general, compliance with official requirements and professional codes of practice is considered high (H. Suter, personal communication, 1997).

Participation in these schemes is voluntary, but the financial incentives to participate are considerable. As a result, the agricultural area farmed according to the principles of integrated production increased from 17% in 1993 to 33% in 1995, and is expected to continue to rise. Participation in organic farming schemes increased from 2.0% of the agricultural area in 1993 to 2.6% in 1995 and an estimated 4.7% in 1996; participation was expected to rise to 10% by 2001 (O. Roux, personal communication, 1997). Reducing the agrochemical input under these extensive production schemes reduces crop yields but, because costs are also lower, profits are maintained.

The reduced use of chemical fertilizers can lead to reduced infestation of the crop by pests and diseases and, traditionally, organic farmers have also used lower sowing rates to reduce the incidence of disease. Swiss farmers have been sowing mixed varieties of cereals, however, which also helps to reduce disease without the use of fungicides. Mixed crops are far less susceptible to the rapid spread of a disease than a monoculture, as the susceptibility of the different varieties to any given disease varies. The success of the mixed-variety approach is underlined by its use elsewhere: Denmark grows more than 60 000 ha of mixed spring barley, and more than 80 000 ha of mixed barley varieties are grown annually in Poland.

England and Wales and France, private companies operate the water industry wholly or in part with government controls to enforce standards. As the need for investment in water services (sewerage and water supply) increases (not only in the eastern part of the European Region but also in the western part) private-sector participation also increases, although only relatively large utilities can

Box 6.4. Agricultural policies to protect groundwater sources in Germany

Federal policy on agriculture in Germany encourages the "extensification" of agriculture, minimal use of chemical products, integrated pest management and organic farming. Legislation and official advice are based on integrated production principles: maximum use of biological defence mechanisms; minimum use of pesticides; and applying the damage threshold principle of using pesticide only if the cost of the potential damage from not applying pesticide exceeds the cost of application.

Protection of groundwater

Environmental protection and protection of groundwater in particular (which provides 85% of the drinking-water in Germany) is characterized by a precautionary approach. The basic expectation is the maintenance of water in its natural state, allowing it to be used as a drinking-water supply without requiring any treatment (groundwater) or minimal treatment only (surface water). The water companies and consumers have an interest in ensuring that abstracted water is of high quality, as this makes expensive, advanced water treatment unnecessary.

Water protection zones around abstraction points have been widely established in Germany to protect groundwater sources used for potable supply. Zone 1 is a 10-metre radius around an abstraction borehole, in which the application of pesticides is banned. Zone 2 is a catchment area defined by 50 days of travel time for the water to reach the abstraction well, and pesticides considered a threat to groundwater (listed in a nationwide federal regulation) are banned in this zone. All the *Länder* must adopt minimum requirements, although individual *Länder* may impose stricter requirements for the various zones.

On a national basis, there has been a move towards protecting all groundwater rather than focusing on protection zones in the catchment areas of wells for drinking-water abstraction. In particular, authorization of new pesticides and the review of previously approved pesticides involve strict assessment in terms of the potential to contaminate groundwater.

Compensation payments and voluntary agreements

In some *Länder*, such as Baden-Württemberg, farmers receive compensation for reduced income caused by the restrictions in the use of agricultural chemicals in water protection zones. Water suppliers pay a charge to the public authorities, who then use the money to provide standard compensation payments to farmers affected by water protection zones.

In other *Länder*, farmers receive compensation for complying with strict environmental criteria, including reducing fertilizer and pesticide use. Special payments are made in sensitive areas for adherence to stricter criteria, such as extensive farming with no fertilizer or pesticide use.

Other schemes to reduce the risk of non-compliance with pesticide standards in drinking-water are also in operation. These include voluntary cooperation between farmers and water suppliers to ensure appropriate management of active ingredients, and some water suppliers employ agricultural engineers to advise farmers on the management necessary to protect water sources. In some cases, compensation for reduced yields (or increased costs for alternative pesticides) may be made on an individual basis after assessment of applications for payments.

Box 6.4. (contd)

The Munich Water Company pays farmers interim compensation payments during a 3-year transition period to organic farming. The Mangfall Valley south-east of Munich was designated as a target area for conversion to organic farming in 1992–1993, to halt the trend of increasing levels of pesticides and nitrate fertilizers in groundwater. By 1996, 70% of the agricultural land in the area was being farmed organically. The high uptake of the scheme is encouraged by the financial support and advice available, but awareness of the importance of ecological soil protection, especially among the younger generation of farmers, is also an important factor (Munich Water Company, unpublished data). A cooperative has been formed to market the organic produce, using promotional slogans such as "one litre of organic milk contributes to protecting 4000 litres of Munich's drinking-water".

Since the scheme started, nitrate levels in groundwater, formerly 14 mg/l, have decreased by about 14%, and pesticides are not detected at levels that would violate national drinking-water standards.

normally attract private investment. When international funding agencies contribute significantly to investment, they also apply pressure to get the services more effectively managed and to impose realistic charges on the users, thereby promoting private investment and operation of the utilities.

The water industry in England and Wales is the only example in Europe of complete privatization. Water service companies cover water supply and sewerage services for entire river basin catchments, except for some smaller water companies operating within these catchments. These companies wholly own the facilities. Local authorities have some involvement in supervising the service and dealing with consumer complaints, but the main supervision is centralized, with the Drinking Water Inspectorate overseeing compliance with quality standards. The companies analyse their own water quality (self-monitoring), but the Drinking Water Inspectorate strictly supervises this.

In France, 50% of wastewater treatment plants are privately operated, and private water companies supply 60% of consumers. With the exception of the Paris water supply company and sewerage services, which the municipal authority wholly owns and manages, the remainder are numerous, mainly small, community services managed directly by local councils or by syndicates of several municipalities. Rural areas still have many individual private suppliers.

Table 6.5. Charges for effluent in selected EU countries

Country	Basis of charges	Comments
England and Wales	Charges reflect monitoring costs, which are related to: • the content of the discharge: for example, monitoring costs are higher for hazardous organic compounds that are more difficult to analyse • the type of recipient water (effluent discharged to vulnerable recipient waters requires more frequent monitoring)	Charges are designed to recover the administration costs of the regulatory authority (Environment Agency of England and Wales) for its pollution control function. The charges are low compared with those in Germany and the Netherlands, and their impact has therefore been relatively low
France	Charges are based on a quantity of pollution produced in a "normal" day during the month in the year with the highest discharge. Each water agency basin committee determines the physico-chemical and biological and microbiological elements to be taken into account. These usually include: • suspended solids • oxidizable matter • toxic substances • phosphorus • nitrogen Charges take into account the vulnerability of the receiving water and the impact of the effluent	These charges implement the "polluter pays" principle but are lower than in Germany and the Netherlands, and their impact has therefore been relatively low. The money raised is used to fund research and infrastructure for pollution abatement
Germany	A national unit charge is applied, independent of the capacity of the recipient water. Charges are reduced by 75% once the limit values laid down in federal regulations for the specific industrial sector have been achieved If the compliance is breached, the reduction in charges no longer applies	Charges can be regarded as an incentive system. The money raised is used to fund research and infrastructures for pollution abatement. The success of the scheme in preventing pollution is probably because of the relatively high charges and the incentive element built into the scheme

Table 6.5. (contd)

Country	Basis of charges	Comments
Netherlands	A unit charge is applied, independent of the capacity of the receiving water	Charges are relatively high, and their impact in improving the environment has therefore been significant. The money raised is used to fund research and infrastructure for pollution abatement

Source: Krinner et al. *(52)*.

Nevertheless, the approach is significantly different from that in England and Wales. In France, the municipality or syndicate owns the facilities, and the ultimate responsibility remains with the local mayor. Time-limited management contracts are issued to private companies for the services. The 95 departmental prefects supervise water quality and technical and administrative matters on behalf of the state, and also monitor water quality or delegate this function.

In most other countries, water services are predominantly organized and managed by local, town or district authorities. There are a variety of approaches, including direct operation and management by local authorities, economic enterprises governed separately by public authorities, and public corporations managed by several municipalities. Some countries increasingly delegate duties to private companies while maintaining public control.

In addition, rural areas still have many very small community or private, individual household supplies with little or no treatment and often minimum supervision of quality.

In the eastern part of the European Region, local authorities operate most of the services, but the major challenge is often much more basic than in the west: continuous supplies of adequate microbial quality. Pressure for private-sector participation is increasing, especially involving water-supply utilities in large towns and cities.

Water quality is supervised in a variety of ways. Self-monitoring as practised in England and Wales and in the Netherlands is relatively rare; government health laboratories or other government-appointed laboratories more typically carry this out. These laboratories are not always the best equipped or resourced to deal with the considerable technical demands of the full range of water quality analyses. For example, the EU Drinking Water Directive *(48)* and the WHO guidelines for drinking-water quality *(46)* prescribe low concentrations of organic contaminants. In addition, the laboratories are often poorly equipped to assess, interpret and report on issues related to access and continuity. These problems have been recognized in some instances. In Germany, for example, the authorities certify some laboratories of water suppliers to analyse compliance on their behalf, but the laboratories are only permitted to monitor the compliance of other water suppliers and not their own.

The organizational arrangements aimed at ensuring compliance with the requirements of legislation, standards or codes of practice must provide for surveillance to be shared between the water-supply agency and a separate, preferably independent, surveillance agency. The water-supply agency carries out routine testing and monitoring (quality control) and the surveillance agency separate checking and testing. Both types of testing should be applied to all types of water available to the community.

The surveillance agency should be established with national support and operate at the central, regional and local levels. It should be concerned with the public health aspects of drinking-water supplies and have overall responsibility for ensuring that all such supplies under its jurisdiction are free from health hazards. To this end, periodic sanitary inspections and analyses of water samples should be carried out to ensure that the suppliers are fulfilling their responsibilities.

Costs and benefits

Many economic consequences follow from the quality of water and sanitation facilities available in a country. Improvements in the level of service will lead to improved health but also involve a cost, as the funds devoted to improving water and sanitation are not available for other purposes. Costs can be viewed from many perspectives, depending on who incurs the costs and who pays them – the people who directly benefit or lose, their families, the health care system or society generally. Reductions in disease produce benefits, primarily but not exclusively by avoiding such costs. Water supply and sanitation services also accrue non-health-related costs and benefits such as convenience and time saved and increased employment.

Whether, to what extent and how rapidly interventions to improve water supply and sanitation conditions should be pursued depends on the costs of such interventions and the consequent benefits, as well as the importance society places on averting avoidable health effects and on the convenience provided by high-quality services. Ignorance of how poor water supply and sanitation systems lead to health and other problems increases the risk of inadequate investment. The balance of costs and benefits differs according to local conditions. One consequence of programmes to control and prevent water-related disease across the European Region could be a narrowing of the gap between countries in the western and eastern parts of the Region. Nevertheless, the western part of the Region still has a significant burden of disease, and this should not be ignored. In Sweden, for example, 90 water-related disease outbreaks, mainly caused by *Campylobacter* spp. and leading to 50 000 cases and two deaths, were reported between 1980 and 1995 *(193)*.

Substantial problems exist in obtaining reliable data, and estimates of costs and benefits are necessarily broad approximations. For the purposes of analysing the issues, the European Region can be split into an eastern and a western part, with the eastern part including the countries of central and eastern Europe and the newly independent states.

Cost models were developed to predict the annual cost of improving water and sanitation service provision to levels consistent with those in the western part of the Region. Cost estimates were based on:

- the existing level of coverage as reported in data on the percentage of the population with a home water supply (access to water within 200 m of the home) and hygienic sanitation (any system that effectively breaks faecal–oral transmission);
- the proportion of existing services that might need to be upgraded;

- the demographic characteristics of the population (degree of urbanization); and
- the characteristics of the investment needed, including level of treatment and other factors.

These calculations are necessarily coarse approximations and are likely to overestimate actual costs because they are engineering-based calculations that ignore more cost-effective local solutions. They are also supply-side costs and ignore what can be achieved by more effectively managing water resources and improving health education. They are blanket measures that ignore cost savings that may be achieved through targeting. The estimated annual costs for the eastern part of the WHO European Region are about €30–50 per person. In all regions except the central Asian republics, these costs are a small proportion (1–2%) of gross domestic product.

From a socioeconomic perspective, the net costs (costs net of any subsequent benefits) are important. The above gross costs overestimate the actual societal costs of the interventions in several ways. First, from a societal perspective they are relevant only if they reflect opportunity costs. The degree of under-utilization of resources in the eastern part of the Region (especially labour) means that the opportunity costs are lower. Second, benefits (some of which can be quantified) offset the estimated costs. The quantifiable benefits include a reduced burden of disease to individuals and society. From a societal perspective, the health benefits need to be valued at what society is willing to pay to secure them (generally measured as the sum of individual willingness to pay). In the absence of primary research on the value society attaches to reducing the level of water-related disease, less accurate empirical methods need to be employed. The most common form of assessment is the cost-of-illness approach. This method attempts to assess the economic implications of disease morbidity and mortality in terms of the costs of the resources lost or required to deal with the disease. Table 6.6 summarizes information from over 25 studies on the economic burden associated with common water-related diseases.

The studies vary widely in their perspectives, many being based on the evaluation of immunization efficacy or on post-outbreak evaluation of resource costs. Few of the studies are based directly on collated

Table 6.6. Costs of illness (morbidity) of water-related diseases

Disease	Value[a]	Coverage[b]	Comments
Hepatitis A			
Severo et al. *(194)*	1 432 (0.079)	1, 2	Prophylaxis for military, hospital workers and tourists in France
Sander et al. *(195)*	9 765 (0.354)	1, 2	Study of hepatitis outbreak in Germany
Arnal et al. *(196)*	909 (0.072)	1	Study of vaccination efficacy in Spain
Verma & Srivastava *(197)*	31 (0.095)	1, 2	Water-related infectious hepatitis in rural populations in India
Demicheli et al. *(198)*	7 109 (0.423)	1, 2	Analysis of an outbreak in Italy mainly affecting children
Lucioni et al. *(199)*	3 735 (0.217)	1, 2, 3, 4	Study of a seafood-based outbreak in Italy
Chossegros et al. *(200)*	1 779 (0.090)	1, 2	Study of cases recorded by hospitals in France
Smith et al. *(201)*	7 310 (0.290)	1, 2	Study of health care workers in the United States
Behrens & Roberts *(202)*	16 170 (1.316)	1, 2, 3, 4	Study on travel prophylaxis in the United Kingdom
Dalton et al. *(203)*	1 577 (0.082)	1, 2, 3, 4	Foodborne outbreak in the United States
Van Doorslaer et al. *(204)*	4 362 (0.343)	1, 2	Travel prophylaxis study in the United Kingdom
Typhoid fever			
Sridhar & Kulkarni *(205)*	22 (0.067)	1	Cost–effectiveness study of antibiotic therapy in India
Shandera et al. *(206)*	4 479 (0.244)	1, 2, 4	Foodborne outbreak of typhoid fever in the United States
Behrens & Roberts *(202)*	9 397 (0.807)	1, 2, 3, 4	Study on travel prophylaxis in the United Kingdom
Diarrhoea			
Archer & Kvenberg *(207)*	2 138 (0.091)	1, 2	Foodborne diarrhoeal disease in the United States
Thomson & Booth *(208)*	19 (0.002)	1, 2	Traveller's diarrhoea in the United Kingdom
Verma & Srivastava *(197)*	10 (0.031)	1, 2	Water-related acute diarrhoeal disease in rural population in India

Table 6.6. (contd)

Disease	Value[a]	Coverage[b]	Comments
Danzon et al. *(209)*	379 (0.020)	1	Efficacy of treatment of children for non-severe acute diarrhoea in France
Gastroenteritis			
Baker et al. *(210)*	94 (0.002)	1, 2, 4	Waterborne outbreak in the United States
Hopkins et al. *(211)*	87 (0.005)	1, 2, 4	Waterborne nonbacterial outbreak in the United States
Liddle et al. *(212)*	896 (0.050)	1, 3, 4	Rotovirus affecting children in Australia
Campylobacteriosis			
Andersson et al. *(193)*	221 (0.009)	1, 2, 3, 4	Study of a waterborne outbreak in Sweden
Laursen et al. *(213)*	104 (0.005)	2	Water-related outbreak in Denmark
Cholera			
Cvjetanovic *(214)*	39 (0.030)	1	Cost benefit of sanitation intervention in Yugoslavia
Cookson et al. *(215)*	761 (0.297)	1	Study of cases in rural Argentina
Enteric fever			
Verma & Srivastava *(197)*	10 (0.030)	1, 2	Study of water-related outbreak in Indian rural populations
Giardiasis			
Harrington et al. *(216)*	5 297 (0.230)	1, 2, 3, 4	Study of waterborne outbreak in the United States
Conjunctivitis			
Verma & Srivastava *(197)*	10 (0.031)	1, 2	Study of water-related acute diarrhoeal disease in rural populations in India
Scabies			
Verma & Srivastava *(197)*	22 (0.068)	1, 2	Study of acute diarrhoeal disease in rural populations in India
Cryptosporidiosis			
Shaffer et al. *(217)*	76 (0.004)	1	Waterborne outbreak in the United States

[a] Costs are absolute costs per person adjusted to 1995 equivalents in euros. The figure in parentheses is the proportion of gross domestic product per person in the study country in question.

[b] Costs included: 1 = health care expenses for patient and society; 2 = direct productivity loss from sufferers; 3: = indirect productivity loss (family care etc. provided to sufferers); 4 = other costs such as purchasing bottled water.

information on costs from individuals experiencing the disease. The studies differ substantially in the costs examined – many are restricted to health care costs and few treat non-marketed labour satisfactorily.

Despite these deficiencies, however, the studies do indicate the relative burden of the morbidity associated with water-related disease. For the more severe diseases (hepatitis A, typhoid fever and cholera) the costs are high at about 10–40% of gross domestic product per person. For less severe diseases, such as diarrhoea and gastroenteritis, the costs are much lower – generally less than 5% of gross domestic product per person.

Data on the incidence of water-related disease were compiled from questionnaires completed by the countries in the WHO European Region. Under-reporting is a significant problem because of poor surveillance systems and, for the less acute diseases, the high number of cases that do not involve the health care services and hence go unreported. These data, adjusted to reflect unreported cases, have been combined with the estimates of the percentage of cases that are considered avoidable through water and sanitation improvements, to estimate the number of avoidable cases in each area in the eastern part of the European Region. Estimates of the excess disease associated with poor water supply and sanitation conditions must recognize the multiple pathways associated with water-related disease. Given the multiple infection pathway, an estimated 60–80% of waterborne disease is avoidable by improving water and sanitation.

These data indicate that more than 30 million cases of water-related disease could be avoided annually through water and sanitation interventions. Based on the cost-of-illness calculations, the economic burden is estimated at about €25 per person in the eastern part of the European Region. These are the benefits associated with avoiding a small subset of all water-related disease and do not include the benefits associated with other important water-related problems: hygiene-related diseases such as scabies, pediculosis and conjunctivitis; skin infections, dermatophytes, etc.; and chemical contaminants such as nitrate, arsenic and lead.

Avoided costs of illness are, however, only one aspect of the health benefits of water-supply and sanitation interventions. The estimated costs of illness ignore the social factors associated with the disease. In addition, improvements in the health status of the general population are likely to positively alter economic conditions at the macroeconomic level: for example, by improving educational attendance, future human capital and savings and investment rates, and by attracting tourism and investment. Specific disease outbreaks can have important macroeconomic effects. In the Peru cholera epidemic in 1991, for example, the lost income from trade and tourism comprised an estimated 34% of the total costs *(218)*.

Investing in water supply and sanitation will produce benefits over and above those associated directly with health. Individuals also value improvements to the conditions and quality of the water supplied, such as the added convenience of a home water connection and fewer restrictions on water use through more continuous supply. Improvements to sanitation will generate environmental benefits through a cleaner environment – raising the quality of the conditions under which the water is used for applications other than drinking, such as aquaculture, recreation, industrial abstraction and irrigation.

Thus the benefits of improving water supply and sanitation conditions in the eastern part of the European Region are probably immense. Although the gross costs are also high, given the importance of the benefits that have not been estimated and the likely overestimation of costs, on balance such investments are likely to produce substantial net benefits. The benefits are likely to increase where well targeted and locally effective strategies are pursued.

7

International, administrative and legal initiatives

Protocol on Water and Health to the 1992 Convention on the Protection and Use of Transboundary Watercourses and International Lakes

The WHO Regional Office for Europe and the United Nations Economic Commission for Europe, with the involvement of the United Nations Environment Programme and the European Union, developed the Protocol on Water and Health to the 1992 Convention on the Protection and Use of Transboundary Watercourses and International Lakes *(3)*.

The Protocol is based on the aims of universal access to drinking-water and sanitation protective of human health and the environment, and of sustainable use of water resources. The aims of the Protocol are:

- universal access to drinking-water
- provision of sanitation for everyone
- sustainable use of water resources.

To achieve these aims, countries should establish, publish and periodically revise national or local targets, ensuring public participation and taking account of: relevant recommendations of international bodies; national and local capacities and resources; and available knowledge related to priorities for the improvement of public health and the protection of the environment, concerning:

- the quality of the drinking-water supplied;
- the area of their territory, or the size or proportion of the population, that should be served by collective systems for the supply of drinking-water, or where the supply of drinking-water by other means should be improved;
- the area of their territory, or the size or proportion of the population, that should be served by collective systems of sanitation, or where sanitation by other means should be improved;
- the standards of performance to be achieved by these systems and by other means of water supply and sanitation;
- the application of recognized good practice to the management of water supply and sanitation;
- the quality of discharge of wastewater to waters within the scope of the Protocol from wastewater collection systems and wastewater treatment installations;
- the disposal or reuse of sewage sludge from collective systems of sanitation or other sanitation installations;
- the quality of waters used as a source of drinking-water or for bathing, irrigation, the production of fish by aquaculture and the production or harvesting of shellfish;
- the application of recognized good practice to the management of enclosed waters generally available for bathing; and
- the performance of systems for the management and protection of water resources, including the application of recognized good practice to the control of pollution from sources of all kinds.

International cooperation will be necessary in many cases, and parties to the Protocol will be required to assist each other in implementing national and local plans. Unexpected events may influence the quality and quantity of water, and precautionary measures and emergency responses should be considered, including:

- improving the security of water supply by abstraction from different water bodies if possible;
- making connections to alternative water supply systems if possible;
- ensuring the availability of equipment for water purification and for the transport of drinking-water from unaffected areas;

- ensuring the availability of equipment to assess the quality of drinking-water; and
- promoting the storage of bottled water in households.

NATIONAL FRAMEWORKS

National legislation and institutional frameworks vary across the WHO European Region and have changed significantly in the last decade in some countries.

European Union countries

In most EU countries, national ministries (often the ministry of environment or ministry of health) are ultimately responsible for enforcing the legislation on water resources and supply. Responsibility may be shared by two ministries, each covering different aspects of water supply. For example, the ministry of the environment may be responsible for water resources (quality and quantity), abstraction licences and discharge permits, while the ministry of health or of foods may take responsibility for drinking-water, bathing water and irrigation water. In some federal countries, however, the responsibilities are divided between the federal level and the individual states, such as in Germany. In practice, supervision of compliance is often fragmented, with much of the responsibility delegated to the regional level and more frequently to the local level.

There are some exceptions, however: notably England and Wales, Ireland and the Netherlands, which have centralized controls on water supply and quality. For England and Wales, where the water industry is fully privatized, the Drinking Water Inspectorate supervises compliance with quality standards of all water companies, the Office of Water Services is in charge of the economic/financial aspects and consumer interests, and the Department of Environment, Transport and the Regions and the Environment Agency of England and Wales (with regional offices based on catchment areas) are responsible for permits for abstraction and discharge. Similar provisions have recently been adopted in Ireland to supervise publicly owned water suppliers. Similarly, in the Netherlands, although somewhat less centralized in practice, regional public health inspectors supervise all water service utilities in their respective regions, and one of the inspectors acts as national coordinator to ensure consistency of approach throughout the regions.

The EU Drinking Water Directive *(47,48)* provides the framework for drinking-water quality standards in the EU. The first Directive was issued in 1980, with member states required to transpose it into national legislation by 1986. They were required to transpose the new Directive into national legislation by 25 December 2000. Issues of the quality of water resources are covered by a variety of directives (and corresponding national legislation). Some of these have been incorporated into the Water Framework Directive, which will aim to establish a framework for protecting and managing water resources.

Although transposing directives into national legislation and subsequent implementation, organizing drinking-water supplies and implementing institutional mechanisms to control compliance with the standards vary considerably in the different EU members, there is a common goal to achieve certain standards. The WHO guidelines for drinking-water quality *(46)* are used as the scientific point of departure in setting individual standards.

All current EU members have transposed the 1980 Drinking Water Directive into national legislation (some albeit with considerable delay) and other relatively new members, such as Austria, are still adjusting to the EU requirements. On the whole, the EU standards have been adopted with minor variation and some additional or stricter national standards in some countries. Many countries still experience problems with certain parameters and are still not fully complying with monitoring requirements and reporting.

Although the system for monitoring and enforcement of compliance in England and Wales has proved effective in raising compliance with drinking-water quality standards, the European Commission has recently challenged the legality of (and started court action on) the approach taken by the Drinking Water Inspectorate in cases where water companies breach standards without compromising public health. The approach involves the use of legally binding undertakings, which are improvement programmes agreed between the Drinking Water Inspectorate and water companies, with clear targets and time limits for achieving these. Many EU members take similar but less formal approaches to enforcement or have, in the past, issued legislation permitting standards to be exceeded temporarily (such as France and Italy); this clearly contravenes the Drinking Water Directive. The Netherlands takes an approach very similar to the enforcement practice of the Drinking Water Inspectorate. Except for one private operator, water services are directly under public control; compliance controls are also carried out through self-monitoring, and the approach to enforcement action taken by public health inspectors is much more informal than in England and Wales but equally successful in achieving improved compliance. Ironically, the revised EU Drinking Water Directive *(48)* now sanctions an approach similar to that currently practised in England and Wales and in the Netherlands.

Gathering and collating information on compliance with the EU Drinking Water Directive are difficult, since few countries publish detailed annual reports on drinking-water quality. Even where reports are produced or data are otherwise available, comparing compliance precisely is difficult because data are presented differently: for example, data may be presented in terms of the percentage of samples analysed, the population or volume of water supplied, supply regions, or compliance with individual parameters or with all parameters measured.

On the whole, supply is continuous in the EU, except for certain localized, seasonal problems resulting from drought. The most frequently reported compliance problems in water quality relate to microbial parameters, nitrate, pesticides and sometimes toxic metals. Similar but more severe problems are encountered in countries in central Europe and particularly in eastern Europe, where supply is still frequently interrupted.

Consumers in France have successfully sued and have been awarded compensation from water suppliers breaching water quality standards. Moreover, France's own public auditing office, the Court des Comptes, in a report in 1997 on water service management, criticized municipalities, utility companies and the relevant state agencies for a lack of transparency, insufficient competition between private operators, inadequate information for consumers and a lack of monitoring of delegated public services.

Enforcement has exerted significant pressure towards implementation of environmental measures, such as stricter controls on pesticide application and designation of water protection zones. This approach has undoubtedly improved the environment in the EU at considerable expense. In addition, expensive treatment technology is frequently needed to remove traces of pesticides from drinking-water at great expense to consumers, while conferring dubious benefits in terms of health effects. The fairness of this approach in terms of violating the polluter-pays principle has often been criticized.

Such debate is particularly pertinent in countries (such as many in the eastern part of the WHO European Region) where continuity of supply and microbial quality are the prime concerns. Considerable care should be exercised in giving priority to investment primarily aimed at securing continuous, safe supplies while, in the short term, avoiding undue emphasis on complying with parameters with little health significance.

Pollution of water sources may in time be affected by the EU Water Framework Directive. In the short term, however, costly treatment options and investment in distribution facilities, as well as effective controls, are widely needed to provide a safe and uninterrupted supply of drinking-water. This results in pressure to attract investment through privatization, which in turn reinforces the need for effective controls to adequately supervise such private operators. These include strong legislation and institutional mechanisms backed by staff training, and adequate resources to allow effective monitoring of drinking-water quality and enforcement of compliance with quality standards. At the same time, overuse of legislation and prosecution of water suppliers for breaching standards can be very costly – ultimately to the consumers – while being likely to provide few or no public health benefits.

Other western European countries

Western European countries that are not members of the EU, such as Norway and Switzerland, have similar national frameworks and legislation in place, with drinking-water quality standards based on a combination of EU limits and WHO guideline values. As these countries are not subject to the legal requirements of the EU, the approach tends to be more pragmatic and focused on health-based criteria.

Switzerland, for example, has two sets of standards: mandatory health-based limits, and non-health-based guideline or target values that are less strictly enforced but ultimately to be aimed towards.

Countries in the eastern part of the WHO European Region

Countries in the eastern part of the WHO European Region have experienced many changes in the past decade. Previous legislation in the USSR and other countries contained numerous strict standards for drinking-water quality, but these were often poorly enforced because institutional mechanisms and resources were lacking. (A notable exception to this was Czechoslovakia.) For most parameters, methods of analysis were inadequate and there was no clear distinction between those responsible for providing services and those responsible for supervising standards and enforcing the law. Moreover, drinking-water quality was often severely compromised by intensive industrial activity without any concern for the environment and, consequently, inadequate protection of water resources. A lack of investment in treatment and distribution facilities also contributed to significant problems.

Many of these countries are preparing or have recently introduced new legislation closely linked to the WHO guidelines and/or the EU Drinking Water Directive, especially candidate countries for membership of the EU. Although some countries, such as the Czech Republic, inherited a relatively sound system of legislation and enforcement, including publication of reports on drinking-water quality, the Czech Republic delayed adopting its revised draft legislation pending adoption of the revised Drinking Water Directive and progress with the Water Framework Directive. Meanwhile, there are concerns that the legislation and enforcement mechanisms have a loophole that could be exploited, especially by newly emerging, inexperienced private water service operators.

As with the EU member states, ministries are responsible for overall water supply in these countries. Although suitable legislation is being prepared and, to some extent, responsibilities are being allocated in most countries, resources and experience are often lacking to enforce the standards effectively. Many countries require considerable investment to improve the infrastructure.

LEGISLATION AND GUIDELINES

Despite the fact that legislation on safe water quality is in place in all WHO European Member States, as well as basic rules on water resource management, there is still an incredible lack of implementation resulting from inconsistencies between legal systems. Successfully conducted case studies on water resource management have shown that space for manoeuvring within the regulations is a key prerequisite for achieving standards above the minimum legal requirements without increasing costs above those that consumers are willing to pay.

EU directives

Nitrates Directive

EU members are required, under the Nitrates Directive *(219)*, to identify bodies of water that may be affected by pollution from nitrate (vulnerable zones) and establish action programmes to prevent pollution in these areas. The Directive is intended both to safeguard drinking-water supplies and to prevent ecological damage, by reducing or preventing the pollution of water caused by the application and storage of inorganic fertilizers and manure on farmland.

Waters covered by the Directive include surface bodies of fresh water (in particular those used for the abstraction of drinking-water), groundwater actually or potentially containing more than 50 mg/l nitrate, and water bodies (lakes, other freshwater bodies, estuaries, coastal water and marine waters) that are, or may become, eutrophic. Action programmes must include periods when applying certain types of fertilizer is prohibited, limits on the quantities of fertilizer applied, a limit on the application of livestock manure, conditions relating to the available storage capacity on farms for livestock manure, and a code of good agricultural practice.

Dangerous Substances Directive

The Dangerous Substances Directive *(220)* sets a framework for eliminating or reducing pollution of inland, coastal and territorial waters by particularly dangerous substances. The regulation of specific substances is promulgated in daughter directives.

The Directive requires EU members to eliminate or reduce pollution of water bodies by certain substances contained in an annex to the directive, and to set standards for their occurrence in water. The dangerous substances to be controlled are contained in two lists, List I (the "black list") of priority chemicals and List II (the "grey list"). Discharges to water of substances on either list must be authorized prior to release. Procedures for determining acceptable levels of release differ between the two lists. List I chemicals are controlled by community-wide emission standards specified in daughter directives, whereas individual EU members are responsible for setting standards for the List II substances that require control.

Groundwater Directive

The Groundwater Directive *(221)* aims to protect exploitable groundwater sources by prohibiting or regulating direct and indirect discharges of dangerous substances. The dangerous substances covered by the directive are those controlled by the Dangerous Substances Directive *(220)*. Member States are required to prevent the introduction of List I substances into groundwater and to limit the introduction of List II substances.

Bathing Water Directive

The Bathing Water Directive (176) sets out the quality requirements for identified bathing waters in each EU member country to "reduce the pollution of bathing water and to protect such water against further deterioration". The standards were set in order "to protect the environment and public health". The Directive specifies minimum sampling frequencies – every two weeks for most parameters. However, the current mandatory and guideline standards of microbial determinants set in the Directive were published before many of the major epidemiological studies had been carried out. The water quality standards of the 1976 Bathing Water Directive are currently being revised after a considerable period of consultation.

Urban Wastewater Treatment Directive

The Urban Wastewater Treatment Directive *(60)* sets minimum standards for the collection, treatment and discharge of urban wastewater, with the aim of reducing the pollution of raw water by domestic sewage, industrial wastewater and rainwater run-off. It introduces controls over the disposal of sewage sludge and prohibits the practice of dumping sewage sludge at sea.

Under the Directive, all towns and villages with a population of 2000 or more are required to have sewage collection systems. The wastewater is subject to treatment requirements; a minimum of secondary treatment is normally required. Tertiary treatment is required for discharge to particularly sensitive areas (as designated by Member States), including waters subject to eutrophication and surface waters intended for abstraction for drinking-water that have high nitrate levels. Exceptions and derogation are made for specific circumstances; for example, septic tanks giving the same degree of protection as sewage collection may be used if the installation of sewerage systems involves "excessive cost".

The Directive requires that all discharges of industrial wastewater into collecting systems and treatment plants be subject to regulation or specific authorization, and is being implemented progressively until 2005.

Drinking-water supply

The EU sets minimum standards for drinking-water quality, supported by monitoring and legal enforcement, and regulations also govern the quality of surface waters abstracted for potable supply and the extent of treatment required.

Such regulations would appear to be necessary to ensure that drinking-water is of acceptable quality and that suitable sources of water are used and sufficient treatment is applied. Nevertheless, the disadvantage of this type of legislation and the concomitant financial penalties for supply companies that breach the quality requirements is that the consumer, and not the polluter, ultimately pays for the treatment to remove the pollution.

Drinking Water Directive

The Drinking Water Directive *(48)* specifies quality standards for water intended for drinking and use in food or drink production. Standards are set for six different categories of parameter:

- organoleptic quality (such as colour, odour and taste) and physicochemical parameters (such as pH and conductivity);
- parameters concerning substances undesirable in excessive amounts (such as nitrates and nitrites);
- toxic substances (such as mercury, lead and pesticides);
- microbial contaminants (such as coliform bacteria and faecal streptococci); and
- the minimum required concentrations for softened water intended for human consumption (such as hardness and alkalinity).

The Directive sets maximum admissible concentrations and minimum required levels for most parameters, which must be incorporated into the legislation of EU members, and includes guidelines for other parameters. The standards are backed up by monitoring and legal enforcement, and regulations also govern the quality of surface water abstracted for potable supply and the extent of treatment required.

The Directive has recently been reviewed and updated. The number of parameters to be regulated were reduced, only those considered to indicate a significant risk to human health being specified.

Surface Water for Drinking Directive

The Surface Water for Drinking Directive *(222)* is intended to ensure that surface water abstracted for use as drinking-water reaches certain standards and receives adequate treatment before being put into public supply. It requires the classification of rivers based on quality (A1, A2 or A3) corresponding to the degree of treatment required to render the surface water fit for supply. Physical, chemical and microbial characteristics are used to define the quality of the water. The 46 parameters include temperature, the five-day BOD test (BOD_5), nitrates, lead and faecal coliform bacteria. Sampling at abstraction points must demonstrate a high degree of compliance with the values required.

The Directive prohibits the use of surface water of a quality worse than A3 for drinking-water except in exceptional circumstances, and requires a plan of action for the improvement of surface water quality, especially A3 water.

WHO GUIDELINES

WHO guidelines for the safe use of wastewater and excreta in agriculture and aquaculture

The WHO guidelines for the safe use of wastewater and excreta in agriculture and aquaculture *(57)* were developed to protect the health of both workers and consumers. They specify that wastewater should be treated to attain certain microbial standards before use, and differentiate between the more stringent quality required for restricted use (on edible crops, sports fields and public parks) and that acceptable for unrestricted use (irrigation of trees, fodder and industrial crops, fruit trees and pasture). Excreta that have not received sufficient treatment to remove the risk of infection should only be applied by subsurface injection or in covered trenches before the start of the growing season. Where waste is used as a nutrient in aquaculture, measures should be implemented to reduce the risk to consumers of fish. These include keeping fish in clean water for at least 2–3 weeks before harvest *(57)*. Plans are being made to update these guidelines.

WHO guidelines for safe recreational-water environments

WHO is developing guidelines for safe recreational-water environments with the primary aim of protecting public health. The guidelines are in two volumes – the first on coastal and fresh waters *(157)* and the second on swimming pools, spas and similar recreational-water environments. Faecal pollution of recreational waters is one of the major hazards facing users, although microbial contamination from other sources as well as chemical and physical aspects also affect the suitability of water for recreation. The guidelines do not provide mandatory limits but recommend measuring the safety of a recreational water environment and promote the adoption of a risk–benefit approach. This approach can then lead to the adoption of measurable standards that can be implemented and enforced, for example to deal with water quality.

WHO guidelines for drinking-water quality

WHO has derived guideline values for a large number of drinking-water parameters *(46)*. These include measures of microbial and chemical contamination and also organoleptic quality. A guideline value for a chemical parameter represents the concentration of a constituent that does not result in any significant risk to the health of the consumer over a lifetime of consumption. The guidelines are intended to set achievable goals that can be used as a basis for the development of national standards that will ensure the safety of drinking-water supplies. They are not intended to be mandatory and should be considered in the context of local or national environmental, social, economic and cultural conditions. The guidelines are reconsidered periodically in the light of new evidence and may be revised if necessary. The ability to achieve the recommended levels is taken into account when setting guidelines.

The guidelines emphasize the overriding importance of ensuring that drinking-water supplies are protected from microbial contamination. The potential consequences of microbial contamination of water supplies are severe, including the simultaneous infection of a large proportion of the population, especially infants and young children, elderly people and people already debilitated by illness. Because chemical contamination is not normally associated with acute effects (except in cases of massive contamination of supplies), chemical standards for drinking-water may be of secondary consideration in a supply subject to severe microbial contamination. Similarly, the risks to health from disinfection by-products are extremely small in comparison with those associated with inadequate disinfection of microbially contaminated supplies.

The guidelines are supported by a series of documents relating to good practice and generally oriented towards specific health hazards or management issues. These documents include a guide to monitoring bathing waters *(177)* and an authoritative review concerning cyanobacteria *(37)*.

International initiatives and responses

World Health Organization

WHO has undertaken several initiatives to improve the health of the population of Europe and to contribute to ensuring that drinking-water and recreational water are of sufficient quality. These include the global guidelines described in the previous section and strategies for health for all.

Health for all

The WHO health for all movement began with the International Conference on Primary Health Care, held in Alma-Ata in 1978 *(223)*. The Member States of the WHO European Region adopted the first set of European targets for health in 1984 *(224)*, updated them in 1991 *(225)* and adopted HEALTH21, the health for all policy framework for the WHO European Region, in 1998 *(4)*. The targets of HEALTH21 include all aspects of health, from achieving environments and lifestyles that are conducive to good health to providing health care.

Target 10 on a healthy and safe physical environment states:

> By the year 2015, people in the Region should live in a safer physical environment, with exposure to contaminants hazardous to health at levels not exceeding internationally agreed standards. In particular: population exposure to physical, microbial and chemical contaminants in water, air, waste and soil that are hazardous to health should be substantially reduced, according to the timetable and reduction rates stated in national environment and health action plans; [and] people should have universal access to sufficient quantities of drinking-water of a satisfactory quality.

This target can be achieved if:

- steps are taken to ensure supply to every home of drinking-water that meets WHO guideline standards for drinking-water quality;
- global water management practices, including pollution control measures, are strengthened;
- proper wastewater management systems are provided for the collection, treatment and final disposal or re-use of all wastewater;
- international conventions, such as those on transboundary waters, are implemented;
- adequate capacities are built up for the inspection and monitoring of health risks in the environment; and
- data collection and monitoring of environmental contamination and health effects are undertaken on a regular basis and their results made freely available.

The problems highlighted in HEALTH21 (underinvestment, poor management of water leading to water shortages, poor management of sanitation and waste services, inadequate treatment of sewage, contamination of water by waste and (particularly in rural areas) by pesticides and nitrates, leakage from drinking-water supplies, interrupted supplies and inefficient use of water supplies) are still very much present in some areas of the European Region.

Although improvements have been made in some aspects and some countries, many of the suggested solutions are as relevant today as they were in 1991. These include: investment in infrastructure, particularly in sewage disposal and treatment plants; urban wastewater treatment; protection of water sources from agricultural, community and industrial wastes; fiscal policies to control pollution; and effective water sector legislation. Investment is required not only for national issues but also for the protection of international waters.

The WHO Regional Director for Europe noted in 1993 that the countries of central and eastern Europe faced the most dramatic challenges in achieving health for all, and considered that adopting the WHO health for all targets as national policies would be the most effective means of creating a cohesive framework for development *(225)*. The 21 current health for all targets together comprise the framework for developing national health policies in the European Region.

European Union

European Community policy addressed environmental and health issues related to water at a time when the Treaty of Rome did not include a mandate for such issues. Thus as early as 1973 the European Community adopted its first environmental action programme, to be followed by others. The 1976 Bathing Water Directive *(176)* directly addressed an issue important for human health: setting criteria for bathing water and obliging European Community members to take the necessary action where the criteria had not yet been met. The 1980 Drinking Water Directive *(47)* for the first time established health-related criteria for drinking-water at the tap, thus providing security for the consumer and a basis for technical and financial planning for the water suppliers. This legislation has been complemented by emission-oriented legislation addressing pollution at the source, such as the Urban Wastewater Treatment Directive in 1991 *(60)*, the Nitrates Directive in 1991 *(219)* and the Integrated Pollution and Prevention Control Directive in 1996 *(226)*.

Nevertheless, European water policy and water legislation still lacked an overall coherence that also contributed to the protection of human health.

Following an initiative taken by the European Commission, a major process of restructuring European water policy is under way, the issue to be addressed by the Water Framework Directive *(2)*. The new policy and legislation have as their main objectives:

- expanding the scope of water protection to all waters, including surface water and groundwater, fresh water, estuaries and marine water;
- achieving good status for all waters by a certain deadline;
- water management based on river basins, by applying the combined approach of limit values and environmental quality standards to the control of discharges and by controlling water abstraction from both surface water and groundwater (EU members would designate river basin authorities to administer and implement the proposed directive, and transboundary cooperation will be required in many instances);
- a combined approach of emission limit values and quality standards;
- getting the prices right;
- involving the citizen more closely; and
- streamlining legislation.

The aim of the Water Framework Directive is to provide an overall framework for the management of water, in terms of both quality and quantity, thus enabling an integrated approach to be taken to achieve the objective of sustainable water management. The proposed directive aims to reconcile all human activities within a catchment, using command and control measures, planning and economic instruments.

The proposed directive incorporates the requirements of a number of current directives, including the Groundwater Directive *(221)*, the Surface Water Directive *(222)* and the Dangerous Substances Directive *(220)*, which are likely to be repealed. Some other directives, such as the Urban Wastewater Treatment Directive *(60)*, the Nitrates Directive *(219)*, the Bathing Water Directive *(176)* and the

Integrated Pollution and Prevention Control Directive *(226)* are likely to remain in force and will provide some of the tools required to implement the Water Framework Directive.

Current legislation on water is often fragmented and inconsistent. Pursuing an integrated policy on water resources and quality management is essential. Sustainable improvements in protecting human health can only be achieved if policies on drinking-water quality take account of wider issues of pollution control, water resource management and social planning.

There is a need to develop and expand local management at the appropriate level, with effective communication at the policy-making level. Intersectoral cooperation is needed in planning, infrastructure, agricultural practices and pollution control *(87)*.

Management of surface water and wastewater quality is of increasing importance in protecting human health. Management tools such as guidelines, quality objectives, discharge permits and cost recovery options must be developed in an appropriate institutional framework.

References

1. EUROPEAN ENVIRONMENT AGENCY. *Europe's environment. The second assessment.* Amsterdam, Elsevier Science, 1998.
2. Directive 2000/60/EC of the European Parliament and of the Council of 23 October 2000 establishing a framework for Community action in the field of water policy. *Official journal of the European Communities*, **43**(L 327): 1–72 (2000).
3. *Protocol on Water and Health to the 1992 Convention on the Protection and Use of Transboundary Watercourses and International Lakes* (http://www.who.dk/london99/water02e.htm). Copenhagen, WHO Regional Office for Europe, 1999 (accessed 1 February 2000).
4. *HEALTH21. The health for all policy framework for the WHO European Region.* Copenhagen, WHO Regional Office for Europe, 1999 (European Health for All Series, No. 6).
5. REES, H.G. ET AL. Estimating the renewable water resource. *In*: Rees, H.G. & Cole, G.A., ed. *Estimation of renewable water resources in the European Union.* Wallingford, Institute of Hydrology, 1997, pp. 38–73 (document SUP-COM95, 95/5-441931EN).
6. STANNERS, D. & BOURDEAU, P., ED. *Europe's environment: the Dobrís assessment. An overview.* Copenhagen, European Environment Agency, 1995.
7. CHERNOGAEVA, G.M. ET AL. Water use and the influence of anthropogenic activity. *In*: Kimstach, V. et al., ed. *A water quality assessment of the former Soviet Union.* London, E. & F.N. Spon, 1998, pp. 69–94.

8. Gleick, P.H. An introduction to global fresh water issues. *In*: Gleick, P.H., ed. *Water in crisis – a guide to the world's fresh water resources*. Stockholm, Pacific Institute for Studies in Development, Environment and Security – Stockholm Environment Institute, 1993, p. 473.
9. *State of the environment: country overview – Russia*. Moscow, State Committee of the Russian Federation on Environmental Protection, 1998, pp. 17–20.
10. Shiklomanov, I.A. et al. Natural water resources. *In*: Kimstach, V. et al., ed. *A water quality assessment of the former Soviet Union*. London, E. & F.N. Spon, 1998, pp. 1–23.
11. *OECD environmental data compendium 1997*. Paris, Organisation for Economic Co-operation and Development, 1997.
12. Schoenwiese, C.-D. et al. *Klimatrend Atlas Europa 1891–1990*. Frankfurt-am-Main, ZUF Verlag, 1993.
13. Brazdil, R. et al. Trends in minimum and maximum daily temperatures in central and southeastern Europe. *International journal of climatology*, **16**: 765–782 (1996).
14. Nicholls, N. et al. *Observed climate variability and change*. *In*: Houghton, J.T. et al., ed. *Climate change 1995. The science of climate change. Contribution of Working Group I to the Second Assessment Report of the Intergovernmental Panel on Climate Change*. Cambridge, Cambridge University Press, 1996, pp. 133–192.
15. Onate, J.J. & Pou, A. Temperature variations in Spain since 1901: a preliminary analysis. *International journal of climatology*, **16**: 805–816 (1996).
16. Watson, R.T. et al., ed. *Climate change 1995. Impacts, adaptations and mitigation of climate change: scientific-technical analyses*. Cambridge, New York, Melbourne, Cambridge University Press, 1966.
17. Kovats, S. et al., ed. *Climate change and stratospheric ozone depletion. Early effects on our health in Europe*. Copenhagen, WHO Regional Office for Europe, 2000 (WHO Regional Publications, European Series, No. 88).
18. Beniston, M. et al. An analysis of regional climate change in Switzerland. *Theoretical and applied climatology*, **49**: 135–159 (1994).
19. Arnell, N.W. & Reynard, N.S. The effects of climate change due to global warming on river flows in Great Britain. *Journal of hydrology,* **183**: 397–424 (1996).

20. Cooper, D.M. et al. The effects of climate changes on aquifer storage and river baseflow. *Hydrological sciences journal*, **40**: 615–631 (1995).
21. Hoermann, G. et al. Auswirkungen einer Temperaturerhöhung auf den Wasserhaushalt der Bornhöveder Seenkette. *EcoSys*, **5**: 27–49 (1995).
22. McMichael, A.J. et al., ed. *Climate change and human health. An assessment prepared by a Task Group on behalf of the World Health Organization, the World Meteorological Organization and the United Nations Environment Programme*. Geneva, World Health Organization, 1996 (document WHO/EHG/96.7).
23. Abakumov, V.A. & Talayeva, Y.G. Microbial pollution. *In*: Kimstach, V. et al., ed. *A water quality assessment of the former Soviet Union*. London, E. & F.N. Spon, 1998, pp. 267–292.
24. Danube Programme Co-ordination Unit. *Workshop on Drinking Water Related Environmental Health Aspects in the Catchment Area of the River Danube, Bratislava, Slovak Republic, 16–19 December 1993. Vol. II. Country reports*. WHO Regional Office for Europe, European Centre for Environment and Health and United Nations Development Programme, 1993 .
25. Petrosyan, V.S. et al. Organic pollutants. *In*: Kimstach, V. et al., ed. *A water quality assessment of the former Soviet Union*. London, E. & F.N. Spon, 1998, pp. 211–240.
26. Meybeck, M. Carbon, nitrogen and phosphorus transport by world rivers. *American journal of science*, **282**: 402–450 (1982).
27. *State of the environment: country overview – Belarus*. Minsk, Ministry of Natural Resources and Environmental Protection of Belarus, 1998.
28. Tsirkunov, V.V. Water quality monitoring systems. *In*: Kimstach, V. et al., ed. *A water quality assessment of the former Soviet Union*. London, E. & F.N. Spon, 1998, pp. 95–112.
29. *State of the environment: country overview – Moldova*. Chisinau, Ministry of Environment of the Republic of Moldova, 1998.
30. *State of the environment: country overview – Ukraine*. Kiev, Ministry of the Environment of Ukraine, 1998.
31. Bernes, C. & Grundsten, C. *The environment*. Stockholm, National Atlas of Sweden, 1992.
32. Koussouris, T.S. et al. Water quality evaluation in lakes of Greece. *In*: *Watershed 89. Proceedings of the IAWPRC conference, Guildford, UK, 17–20 April 1989*. Oxford, Pergamon Press, 1989, pp. 119–128.

33. SCHEIDLEDER, A. ET AL. *Groundwater quality and quantity. Data and basic information.* Copenhagen, European Environment Agency, 1999 (Technical Report No. 22).
34. EUROPEAN ENVIRONMENT AGENCY. *Environment in EU at the turn of the century.* Copenhagen, European Environment Agency, 1999.
35. *World register of dams. Complete register.* Paris, International Commission on Large Dams, 1984.
36. *World register of dams. Revision volume.* Paris, International Commission on Large Dams, 1988.
37. CHORUS, I. & BARTRAM, J., ED. *Toxic cyanobacteria in water.* London, E. & F.N. Spon, 1999.
38. KLEIN, G. Application of the OECD eutrophication model in the process of ogliotrophication in eutrophied waters. *Vom Wasser,* **73**: 365–373 (1989).
39. RYDING, S.-O. & RAST, W. *The control of eutrophication of lakes and reservoirs.* New York, Parthenon Publishing Group, 1989.
40. SAS, H. *Lake restoration by reduction of nutrient loading: expectations, experiences, extrapolations.* St Augustin, Academia Verlag, 1989.
41. KLAPPER, H. *Control of eutrophication in inland waters.* Chichester, Ellis Horwood, 1991.
42. *State of the environment: country overview – Georgia.* Tbilisi, Ministry of Environment of Georgia, 1998.
43. MARGAT, J. *L'eau dans le bassin Méditerranéen. Situation et prospective. Les Fascicules du Plan Bleu 6.* Paris, Economica, 1992.
44. MINISTRY OF TRANSPORT AND PUBLIC WORKS. *Report on the Dutch aquatic outlook: future for water.* The Hague, Directorate-General of Public Works and Water Management, 1996.
45. CROOK, J. ET AL. *Guidelines for water reuse.* Cambridge, MA, Camp Dresser & McKee, Inc., 1992.
46. *Guidelines for drinking-water quality,* 2nd ed. Geneva, World Health Organization, 1993–1998.
47. Council directive 80/778/EEC of 15 July 1980 relating to the quality of water intended for human consumption. *Official journal of the European Communities,* **L229**: 11–29 (1980).
48. Council directive 98/83/EC of 3 November 1998 on the quality of water intended for human consumption. *Official journal of the European Communities,* **L330**(05/12/1998): 32–54 (1998) (accessed 5 March 2000).

49. Fielding, M. et al. *Pesticides in groundwater. Final report to European Crop Protection Association*. Medmenham, Water Research Centre, 1998.
50. *Water supply and sanitation in central and east European countries, new independent states and Mongolia*. Hall/Tirol, Mountain Unlimited, 1995.
51. *Convention on the Protection and Use of Transboundary Watercourses and International Lakes*. Geneva, United Nations Economic Commission for Europe, 1992.
52. Krinner, W. et al. *Sustainable water use in Europe. Part 1. Sectoral use of water*. Copenhagen, European Environment Agency, 1999 (Environmental Assessment Report No. 1).
53. Federov, Y.A. et al. The Amu Darya. *In*: Kimstach, V. et al., ed. *A water quality assessment of the former USSR*. London, E. & F.N. Spon, 1998, pp. 413–433.
54. Kohsiek, L.H.M., ed. *Backdrop on the 1995 environmental balance*. Bilthoven, National Institute of Public Health and Environmental protection, 1995 (Report 251701023).
55. Desbrow, C. et al. *The identification and assessment of oestrogenic substances in sewage treatment effluents*. Bristol, Environment Agency, 1996 (Report P2-i490/7).
56. Pucelj, G. Will less technology solve sewage problems? *Danube watch*, **2**(4): 10–11 (1996).
57. *Health guidelines for the use of wastewater in agriculture and aquaculture. Report of a WHO Scientific Group*. Geneva, World Health Organization, 1989 (Technical Report Series, No. 778).
58. Mainstone, C. et al. *The environmental impact of fish farming – a review*. Medmenham, Water Research Centre, 1989 (Report No. PRS 2243-M).
59. Rana, K. & Immink, A. *Trends in global aquaculture production: 1984–1996*. Rome, Food and Agriculture Organization of the United Nations, 1999.
60. Council directive 91/271/EEC of 21 May 1991 concerning urban wastewater treatment. *Official journal of the European Communities*, **L135**: 40–52 (1991).
61. Kauppi, L. et al. Impacts of agricultural nutrient loading on Finnish watercourses. *Water science and technology*, **28**: 461–471 (1993).
62. Sims, J.T. Agricultural and environmental issues in the management of poultry wastes: recent innovations and long-term challenges. *ACS symposium series*, **668**: 72–90 (1997).

63. *FAOSTAT statistics database (agriculture – irrigation)*. Rome, Food and Agriculture Organization of the United Nations, 1996.
64. Hansen, J. *Nitrogen balances in agriculture*. Luxembourg, EUROSTAT, 2000.
65. Sibbesen, E. & Runge-Metzger, A. Phosphorus balance in European agriculture – status and policy options. *SCOPE*, **54**: 43–60 (1995).
66. *Fish farming and the Scottish freshwater environment. Report prepared for the Nature Conservancy Council by the Institute of Aquaculture*. Stirling, University of Stirling, 1990.
67. Foy, R.H. & Rosell, R. Loadings of nitrogen and phosphorus from a Northern Ireland fish farm. *Aquaculture*, **96**: 17–30 (1991).
68. *European crop protection: trends in volumes sold, 1985–95*. Brussels, European Crop Protection Association, 1996.
69. Blackmore, J. & Clark, L. *The disposal of sheep dip waste: effects on water quality*. Bristol, National Rivers Authority, 1994.
70. *Sheep dipping: pollution prevention guidelines*. Bristol, Environment Agency, 1997.
71. Lee, E.M. et al. *The occurrence and significance of erosion, deposition and flooding in Great Britain*. London, H.M. Stationery Office, 1995.
72. Lee, E.M. et al. *The investigation and management of erosion, deposition and flooding in Great Britain*. London, H.M. Stationery Office, 1995.
73. Boardman, J. & Robinson, D.A. Soil erosion, climate vagarity and agricultural change on the Downs around Lewes and Brighton, autumn 1982. *Applied geography*, **5**: 243–258 (1985).
74. Evans, R. & Cook, S. Soil erosion in Britain. *SEESOIL*, **3**: 28–58 (1986).
75. Spiers, R.B. & Frost, C.A. Soil water erosion on arable land in the United Kingdom. *Research and development in agriculture*, **4**: 1–11 (1987).
76. Reed, A.H. Soil loss from tractor wheelings. *Soil and water*, **14**: 12–14 (1986).
77. *State of the environment of the Russian Federation. 1993 national report*. Moscow, Ministry of Environmental Protection and Natural Resources, 1994.
78. Czech Environmental Institute, ed. *Statistical environmental yearbook of the Czech Republic 1996*. Prague, Ministry of the Environment, 1996.

79. Longstaff, S.L. et al. Contamination of the chalk aquifer by chlorinated solvents: a case study of the Luton and Dunstable area. *Journal of the Institute of Water and Environmental Management*, **6**: 541–550 (1992).
80. Cavellero, A. et al. Underground water pollution in Milan by industrial chlorinated organic compounds. *In*: De L.G. Solbe, J.F., ed. *Effects of land use on fresh waters: agriculture, forestry, mineral exploitation, urbanisation*. Chichester, Ellis Horwood, 1985, pp. 65–84.
81. Baxter, K.M. The effects of a hazardous and a domestic waste landfill on the trace organic quality of chalk groundwater at a site in East Anglia. *Science of the total environment*, **47**: 93–98 (1985).
82. Lerner, D.N. & Tellam, J.H. The protection of urban groundwater from pollution. *Journal of the Institution of Water and Environmental Management*, **6**(1): 28–36 (1992).
83. Rivett, M.O. et al. Chlorinated solvents in UK aquifers. *Journal of the Institute of Water and Environmental Management*, **4**: 242–250 (1990).
84. *Proceedings of the European Workshop on the Impact of Endocrine Disrupters on Human Health and Wildlife, Weybridge, 2–4 December 1996*. Copenhagen, European Environment Agency, 1997.
85. Harries, J.E. et al. *Effects of trace organics on fish – phase 2*. Marlow, Foundation for Water Research, 1995 (Report FR/D 0022).
86. James, H.A. et al. *Steroid concentrations in treated sewage effluents and water courses – implications for water supplies*. Medmenham, Water Research Centre, 1998 (Report No. 98 TX011).
87. WHO European Centre for Environment and Health. *Concern for Europe's tomorrow. Health and the environment in the WHO European Region*. Stuttgart, Wissenschaftliche Verlagsgesellschaft, 1995.
88. Somlyódy, L. & Jolánkai, G. Nutrient loads. *In*: Somlyódy, L. & van Straten, G., ed. *Modeling and managing shallow lake eutrophication*. Berlin, Springer, 1986, pp. 125–156.
89. Mara, D. & Cairncross, S. *Guidelines for the safe use of wastewater and excreta in agriculture and aquaculture. Measures for public health protection*. Geneva, World Health Organization, 1989.

90. *Water and sanitation services in Europe, 1981–1990.* Copenhagen, WHO Regional Office for Europe, 1989 (document EUR/ICP/CWS 011).
91. *Environment and health in Italy.* Rome, WHO European Centre for Environment and Health, Rome Division, 1995.
92. BRTKO, J. ET AL. *National report, Slovak Republic.* Bratislava, Water Research Institute, 1997.
93. *Water supply and sanitation in central and east European countries, new independent states and Mongolia. Vol. II, draft version.* Hall/Tirol, Mountain Unlimited, 1997.
94. BERTOLLINI, R. ET AL., ED. *Ambiente e salute in Italia* [Environment and health in Italy]. Rome, Il Pensiero Scientifico Editore, 1997.
95. IACOB, I. A decade of drinking water surveillance in Romania. *In*: *Proceedings of the XXXII International 70th Anniversary Conference of the Institute of Public Health, Bucharest, Romania, May 8–9, 1997.* Bucharest, Institute of Public Health, 1997, pp. 122–123.
96. *Eradication of water-related diseases. Report on a joint UNECE/WHO consultation.* Copenhagen, WHO Regional Office for Europe, 1997 (document EUR/ICP/EHSA 02 02 03).
97. PALOMBI, L. ET AL. *Tirana survey on water facilities.* Tirana, Community of S. Edido, 1997.
98. *Epidemiological aspects of investigation of outbreaks of communicable diseases and surveillance and control of water quality. Report on a WHO workshop.* Copenhagen, WHO Regional Office for Europe, 1997 (document EUR/ICP/CORD 02 06 01).
99. *International statistics for water supply.* Nancy, International Water Supply Association, 1997.
100. *Financial management of water supply and sanitation. A handbook.* Geneva, World Health Organization, 1994.
101. *Dix ans de croissance pour l'industrie européenne des eaux minérales naturelles.* Madrid, GISEM-UNESEM, 1993.
102. *Joint WHO/German Regional Seminar on Drinking-Water Quality. Report on a WHO seminar.* Copenhagen, WHO Regional Office for Europe, 1994 (document EUR/ICP/EHAZ 94 11/WS03).
103. *Environment for sustainable health development – an action plan for Sweden. Report on the proposals presented by the Commission on Environmental Health.* Stockholm, SOU, 1996.
104. RAUKAS, A. *Estonian environment: past, present and future.* Tallinn, Ministry of the Environment of Estonia, Environment Information Centre, 1996.

105. *Guidelines for drinking-water quality*, 2nd ed. *Vol. 2. Health criteria and supporting information.* Geneva, World Health Organization, 1996.

106. HALL, T., ED. *Water treatment processes and practices*, 2nd ed. Swindon, Water Research Centre, 1997.

107. *Outbreak of typhoid fever in Tajikistan. Report of a WHO expert team visit to Tajikistan, 1–9 July 1996.* Copenhagen, WHO Regional Office for Europe, 1996 *Policy aspects of water-related issues. Report on a WHO consensus meeting.* Copenhagen, WHO Regional Office for Europe, 1997 (document EUR/ICP/CORD 02 06 01(A)).

108. SIBILLE, I. Biological stability in drinking water distribution systems : a review. *L'année biologique*, **37**(3): 117–161 (1998).

109. VAN DER HOEK, J.P. ET AL. RO treatment: selection of a pre-treatment scheme based on fouling characteristics abnd operationg conditions based on environmental impact. *Desalination*, **127**: 89–101 (2000).

110. WATER RESEARCH CENTRE. *International comparison of the demand for water: a comparison of the demand for water in three European countries: England and Wales, France and Germany.* London, Office of Water Services, 1997.

111. ISTITUTO DI RICERCA SULLE ACQUE. *Evoluzione dei fabbisogni idrici civili ed industriali* [Development of human and industrial water requirements]. Cosenza, Editoriale Bios, 1996.

112. DA C. VARGAS, S.M.V. & MARA, D.D. The bacterial quality of lettuce and alfalfa spray-irrigated with trickling filter effluent. *In*: *Implementing water re-use (Water Re-use Symposium IV).* Denver, CO, American Water Works Association Research Foundation, 1987, pp. 793–802.

113. MARA, D.D. & PEARSON, H.W. *Design manual for waste stabilisation ponds in Mediterranean countries.* Leeds, Lagoon Technology International, 1998.

114. *Workshop on Water Demands Management, Frejus, 12–13 September, 1997.* Valbonne, Plan Bleu, 1997.

115. BLACK, R.E. ET AL. Handwashing to prevent diarrhoea in day-care centres. *American journal of epidemiology*, **113**: 445–451 (1981).

116. MCNEISH, D. *Liquid gold. The cost of water in the 90's.* Ilford, Barnardos, 1993.

117. MIDDLETON, J.D. ET AL. Water disconnection and disease. *Lancet*, **344**: 62 (1994).

118. Fewtrell, L. et al. Infectious diseases and water-supply disconnections. *Lancet*, **343**: 1370 (1994).
119. *Annual national report on drinking water quality in urban areas.* Bucharest, Institute of Public Health, 1986–1996.
120. *The quality of drinking water in Ireland. A report for the year 1995 with a review of the period 1993–1995.* Wexford, Environmental Protection Agency, 1996.
121. Maguire, H.C. et al. An outbreak of cryptosporidiosis in South London – what value the *P* value? *Epidemiology and infection*, **115**: 279–287 (1995).
122. Braidech, T.E. & Karlin, R.J. Causes of a waterborne giardiasis outbreak. *Journal of the American Water Works Association*, **77**: 48–51 (1985).
123. Furtado, C. et al. Outbreaks of waterborne infectious disease in England and Wales, 1992–95. *Epidemiology and infection*, **121**: 109–119 (1998).
124. Karanis, P. et al. Distribution and removal of *Giardia* and *Cryptosporidium* in water supplies in Germany. *Water science and technology*, **37**: 9–18 (1998).
125. *Foodborne pathogens: risks and consequences.* Ames, IA, Council for Agricultural Science and Technology, 1994 (Task Force Report No. 122).
126. Helmick, C.G. et al. *Infectious diarrheas. In*: Everhart, J.E., ed. *Digestive diseases in the United States: epidemiology and impact.* Bethesda, MD, National Institutes of Health, 1994, pp. 85–120 (Publication No. 94-1447).
127. Food safety and foodborne diseases. *World health statistics quarterly*, **50**(1/2): 3–154 (1997).
128. Kramer, M.H. et al. Waterborne disease: 1993–1994. *Journal of the American Water Works Association*, **88**(3): 66–80 (1996).
129. MacKenzie, W.R. et al. A massive outbreak in Milwaukee of *Cryptosporidium* infection transmitted through the public water supply. *New England journal of medicine*, **331**: 161–167 (1994).
130. Proctor, M.E. et al. Surveillance data for waterborne illness detection: an assessment following a massive waterborne outbreak of *Cryptosporidium* infection. *Epidemiology and infection*, **120**(1):43–54 (1998).
131. Goldstein, S.T. et al. Cryptosporidiosis: an outbreak associated with drinking water despite state-of-the-art water treatment. *Annals of internal medicine*, **124**: 459–468 (1996).

132. Roefer, P.A. et al. The Las Vegas cryptosporidiosis outbreak. *Journal of the American Water Works Association*, **88**(9): 95–106 (1996).
133. Exner, M. Infektionskrankheiten aus hygienischer Sicht mit besonderer berücksichtigung umweltbedingter Infektionen – Rückblick und Ausblick. *Zentralblatt für Hygiene und Umweltmedizin*, **197**: 134–161 (1995).
134. Payment, P. et al. A prospective epidemiological study of drinking water related gastrointestinal illnesses. *Water science and technology*, **24**(2): 27–28 (1991).
135. Payment, P. et al. A randomized trial to evaluate the risk of gastrointestinal disease due to consumption of drinking water meeting current microbiological standards. *American journal of public health*, **81**: 703–708 (1991).
136. Gray, N.F., ed. *Drinking water quality: problems and solutions.* Chichester, John Wiley & Sons, 1994.
137. Simchen, E. et al. An epidemic of waterborne *Shigella* gastroenteritis in kibbutzim of western Galilee in Israel. *International journal of epidemiology,* **20**: 1081–1087 (1991).
138. Hunter, P.R., ed. *Waterborne disease. Epidemiology and ecology*. Chichester, John Wiley & Sons, 1997.
139. Narkevich, M.I. et al. The seventh pandemic of cholera in the USSR, 1961–1989. *Bulletin of the World Health Organization*, **71**: 189–196 (1993).
140. Grassi, E. et al. The control of cholera epidemics: continuity or discordance in the health interventions. *Igiene moderna*, **109**: 541–556 (1998).
141. Schiraldi, O. Tourist health and enteric infections – cholera. *In*: Pasini, W., ed. *Tourist health. Proceedings of the Second International Conference on Tourist Health, Rimini, 15–18 March 1989*. Rimini, WHO Collaborating Centre for Tourist Health and Tourist Medicine, 1990, pp. 95–97.
142. Smith, H.V. et al. *The effect of free chlorine on the viability of Cryptosporidium spp. oocysts*. Medmenham, Water Research Centre, 1988 (Report PRU 2023-M).
143. Crane, G.F. et al. Waterborne outbreaks of cryptosporidiosis. *Journal of the American Water Works Association*, **90**(9): 81–91 (1998).
144. Richardson, A.J. et al. An outbreak of waterborne cryptosporidiosis in Swindon and Oxfordshire. *Epidemiology and infection*, **107**: 485–495 (1991).

145. Fewtrell, L. & Delahunty, A. The incidence of cryptosporidiosis in comparison with other gastro-intestinal illnesses in Blackpool, Wyre and Fylde. *Journal of the Chartered Institution of Water and Environmental Management*, **9**: 598–601 (1995).

146. Duke, L.A. et al. A mixed outbreak of *Cryptosporidium* and *Campylobacter* infection associated with a private water supply. *Journal of epidemiology and infection*, **116**: 303–308 (1996).

147. Ljungstrøm, I. & Castor, B. Immune response to *Giardia lamblia* in a water-borne outbreak of giardiasis in Sweden. *Journal of medical microbiology*, **36**: 347–352 (1992).

148. Zhulidov, A.V. & Emetz, V.M. Heavy metals, natural variability and anthropogenic impacts. *In*: Kimstach, V. et al., ed. *A water quality assessment of the former Soviet Union*. London, E. & F.N. Spon, 1998, pp. 179–202.

149. *Guidelines for drinking-water quality. Vol. 1*. Geneva, World Health Organization, 1984.

150. Fawell, J.K. et al. *Toxins from blue-green algae: toxicological assessment of microcystin-LR and a method for its determination in water*. Medmenham, Water Research Centre, 1994 (Report FR 0359/ DoE 3358/2).

151. Hart, J. & Stott, P. *Microcystin-LR removal from water*. Marlow, Foundation for Water Research, 1993 (Report FR 0367).

152. Lahti, K. & Hiisvirta, L. Removal of cyanobacterial toxins in water treatment processes: review of studies conducted in Finland. *Water supply*, **7**: 149–154 (1989).

153. Wroath, A. & Fawell, J.K. *The toxicity and significance of toxins from blue-green algae*. Medmenham, Water Research Centre, 1995.

154. Falconer, I.R. Health problems from exposure to cyanobacteria and proposed safety guidelines for drinking and recreational water. *In*: Codd, G.A. et al., ed. *Detection methods for cyanobacterial toxins*. London, Royal Society of Chemistry, 1994, pp. 3–30.

155. Cronberg, G. Phytoplankton changes in Lake Trummen induced by restoration. *Folia limnologica scandinavica*, **18**: 119 (1982).

156. Turner, P.C. et al. Pneumonia associated with cyanobacteria. *British medical journal*, **200**: 1440–1441 (1990).

157. *Guidelines for safe recreational-water environments. Vol. 1. Coastal and fresh-waters*. Geneva, World Health Organization, 1998 (document EOS/DRAFT/98.14).

158. Auvinen, A. et al. Indoor radon exposure and risk of lung cancer: a nested case-control study in Finland. *Journal of the National Cancer Institute*, **88**: 966–972 (1996).
159. *Drinking water quality in Scotland, 1996*. Edinburgh, Scottish Executive, 1997.
160. Martyn, C.N. et al. Geographical relation between Alzheimer's disease and aluminium in drinking-water. *Lancet*, **1**: 59–62 (1989).
161. *Aluminium*. Geneva, World health Organization, 1997 (Environmental Health Criteria, No. 194).
162. Bennet, G. Bristol floods 1968: controlled survey of effects of health on local community disaster. *British medical journal*, **3**: 454–458 (1970).
163. Abrahams, M.J. et al. The Brisbane floods, January 1974: their impact on health. *Medical journal of Australia*, **2**: 936–939 (1976).
164. International Federation of Red Cross and Red Crescent Societies. *World disaster report 1997*. New York, Oxford University Press, 1998.
165. Simoes, J. et al. [Some aspects of Weil's disease epidemiology based on a recent epidemic after a flood in Lisbon (1967).] *Anais da Escola Nacional de Saude Publica e de Medicina Tropical*, **3**: 19–32 (1969).
166. Kriz, B. *Infectious disease consequences of the massive 1997 summer floods in the Czech Republic*. Copenhagen, EWHO Regional Office for Europe, 1998 (document EUR/ICP/EHRO 020502/12).
167. Dales, R.E. et al. Adverse health effects among adults exposed to home dampness and molds. *American review of respiratory diseases*, **143**: 505–509 (1991).
168. Estrela, T. et al. *Water resources problems in southern Europe. An overview report*. Copenhagen, European Environment Agency, 1996.
169. *Microbiological quality of coastal recreational waters. Report on a joint WHO/UNEP meeting*. Copenhagen, WHO Regional Office for Europe, 1994 (document EUR/ICP/CEH/039(1).
170. *World health statistics annual 1995*. Geneva, World Health Organization, 1996.
171. Centers for Disease Control and Prevention. Surveillance for waterborne disease outbreaks – United States, 1995–1996. *Surveillance summaries*, December 11, pp. 1–34 (1999).

172. Pike, E.B. *Health effects of sea bathing*. Medmenham, Water Research Centre, 1994 (Report No. DoE 3412/2).
173. Fewtrell, L. et al. *Pathogenic microorganisms in temperate environmental waters*. Dyfed, Samara Publishing, 1994.
174. Fleisher, J.M. et al. Marine waters contaminated with domestic sewage: nonenteric illnesses associated with bather exposure in the United Kingdom. *American journal of public health*, **86**: 1228–1234 (1996).
175. Pruss, A. A review of epidemiological studies from exposure to recreational water. *International journal of epidemiology*, **27**: 1–9 (1998).
176. Council directive 76/160/EEC of 8 December 1975 concerning the quality of bathing water. *Official journal of the European Communities*, **L031**: 1–7 (1976).
177. Bartram, J. & Rees, G. Approaches to microbiological monitoring. *In*: *Monitoring bathing waters: a practical guide to the design and implementation of assessments and monitoring programmes*. London, E. & F.N. Spon, 2000, pp. 169–218.
178. *Health-based monitoring of recreational waters: the feasibility of a new approach (the "Annapolis Protocol"). Outcome of an expert consultation, Annapolis, USA*. Geneva, World Health Organization, 1999 (document WHO/SDE/WSH/99.1).
179. *Quality of bathing water (1997 bathing season)*. Luxembourg, European Commission, 1998 (EUR 18166).
180. Pilotto, L.S. et al. Health effects of recreational exposure to cyanobacteria (blue-green) algae during recreational water-related activities. *Australia and New Zealand journal of public health*, **21**: 562–566 (1997).
181. Pearson, M.J. et al. *Toxic blue-green algae*. London, National Rivers Authority, 1990 (Water Quality Series, No. 2).
182. Gunn, G. et al. Fatal canine neurotoxicosis attributed to blue-green algae (cyanobacteria). *Veterinary record*, **130**: 301–302 (1992).
183. Grauer, F. Seaweed dermatitis. *Archives of dermatology*, **84**: 720–732 (1961).
184. Baden, D.G. et al. Toxins from Florida's red tide dinoflagellates, *Ptychodiscus brevis*. *In*: Regelis, E.P., ed. *Seafood toxins*. Washington, DC, American Chemical Society, 1984, pp. 359–367.
185. Scoging, A.C. Illness associated with seafood. *Communicable disease report, England and Wales*, **1**: 117–122 (1991).

186. *Assessment of the state of microbiological pollution of the Mediterranean Sea*. Nairobi, United Nations Environment Programme, 1996 (document UNEP(OCA)/MED WG. 104/Inf.9).
187. Blake, P.A. et al. Cholera in Portugal, 1974. I. Modes of transmission. *American journal of epidemiology*, **105**: 337–343 (1977).
188. *Health risks from marine pollution in the Mediterranean. 2. Review of hazards and health risks*. Copenhagen, WHO Regional Office for Europe, 1995 (document EUR/ICP/EHAZ 94 01/MTO1(2)).
189. *Towards sustainability: a European Community programme of policy and action in relation to the environment and sustainable development*. Brussels, Commission of the European Communities, 1992 (COM(92) 23).
190. Buckland, J. & Zabel, T. *Economic instruments of water management and financing of infrastructure*. Bristol, Environment Agency, 1996 (R&D Technical Report E10).
191. Agthe, D.E. & Billings, R.B. Water price effect on residential and apartment low flow fixtures. *Journal of water resources planning and management*, **122**(1): 20–23 (1996).
192. *Long range study on water supply and demand in Europe – level A – France*. London, International Water Association, 1996.
193. Andersson, Y. et al. Waterborne *Campylobacter* in Sweden: the cost of an outbreak. *Water science and technology*, **35**(11–12): 11–14 (1997).
194. Severo, C.-A. et al. Cost effectiveness of hepatitis A prevention in France. *Pharmacoeconomics*, **8**: 46–61 (1995).
195. Sander, G. et al. Zur Problematik der Kostendarstellung übertragbarer Krankheiten. IV. Anwendungsversuch einer Kostenrechnung bei Hepatitis infectiosa am Beispiel einer Hepatitisepidemie im zwei Kreisen. *Zeitschrift für ärztliche Fortbildung (Jena)*, **72**: 194–200 (1978).
196. Arnal, J.-M. et al. Cost effectiveness of hepatitis A virus immunisation in Spain. *Pharmacoeconomics*, **12**: 361–373 (1997).
197. Verma, B.L. & Srivastava, R.N. Measurement of the personal cost of illness due to some major water-related diseases in an Indian rural population. *International journal of epidemiology*, **19**: 169–176 (1990).
198. Demicheli, V. et al. Economic aspects of a small epidemic of hepatitis A in a religious community in northern Italy. *Journal of infection*, **33**: 87–90 (1996).

199. Lucioni, C. et al. Cost of an outbreak of hepatitis A in Puglia, Italy. *Pharmocoeconomics*, **13**: 257–266 (1998).
200. Chossegros, P. et al. Coût en France des hepatites A aïgues de l'adulte. *Presse medicale*, **23**: 561–564 (1994).
201. Smith, S. et al. Cost-effectiveness of hepatitis A vaccination in healthcare workers. *Infection control and hospital epidemiology*, **18**: 688–691 (1997).
202. Behrens, R.H. & Roberts, J.A. Is travel prophylaxis worth while? Economic appraisal of prophylactic measures. *British medical journal*, **309**: 918–922 (1994).
203. Dalton, C.B. et al. The cost of a food-borne outbreak of hepatitis A in Denver, Colorado. *Archives of internal medicine*, **156**: 1013–1016 (1996).
204. Van Doorslaer, E. et al. Cost-effectiveness analysis of vaccination against hepatitis A in travellers. *Journal of medical virology*, **44**(4): 463–469 (1994).
205. Sridhar, C.B. & Kulkarni, R.D. Reassessment of frequency of occurrence of typhoid fever and cost efficacy analysis of antibiotic therapy. *Journal of the Association of Physicians – India*, **43**: 679–684 (1995).
206. Shandera, W.Z. et al. An analysis of the economic costs associated with an outbreak of typhoid fever. *American journal of public health*, **75**: 71–73 (1985).
207. Archer, D.L. & Kvenberg, J.E. Incidence and cost of foodborne diarrhoeal disease in the United States. *Journal of food protection*, **48**: 887–894 (1985).
208. Thomson, M.A. & Booth, I.W. Treatment of traveller's diarrhoea. Economic aspects. *Pharmacoeconomics*, **9**: 382–391 (1996).
209. Danzon, A. et al. Prise en charge des diarrhées aïgues non graves du nourrisson: approche socio-économique. *Revue d'épidémiologie et de santé publique*, **35**: 451–457 (1987).
210. Baker, E.L. Jr. et al. Economic impact of a community wide waterborne outbreak of gastrointestinal illness. *American journal of public health*, **69**: 501–502 (1979).
211. Hopkins, R.S. et al. Gastroenteritis: case study of a Colorado outbreak. *Journal of the American Water Works Association*, **78**(1): 40–44 (1986).
212. Liddle, J.L. et al. Rotavirus gastroenteritis: impact on young children, their families and the health care system. *Medical journal of Australia*, **167**: 304–307 (1997).

213. LAURSEN, E. ET AL. Gastroenteritis: a waterborne outbreak affecting 1600 people in a small Danish town. *Journal of epidemiology and community health*, **10**: 453–458 (1994).
214. CVJETANOVIC, B. Sanitation versus immunization in the control of enteric and diarrhoeal diseases. *Progress in water technology*, **11**: 81–87 (1979).
215. COOKSON, S.T. ET AL. A cost-benefit analysis of programmatic use of CVD 103-HgR live oral cholera vaccine in a high risk population. *International journal of epidemiology*, **26**: 212–219 (1997).
216. HARRINGTON, W. ET AL. The economic losses of a waterborne disease outbreak. *Journal of urban economics*, **25**: 116–138 (1989).
217. SHAFFER, P.A. ET AL. *The impact of a community-wide waterborne outbreak of cryptosporidiosis on local health-care resources* (abstract). *In*: *Abstracts of the 13th AHSR-FHSR Annual Meeting, 1996*. Washington, DC, AHSR-FHSR, 1996, p. 99.
218. The economic impact of the cholera epidemic, Peru, 1991. *Epidemiology bulletin*, **13**(3): 9–11 (1992).
219. Council directive 91/676/EEC of 12 December 1991 concerning the protection of waters against pollution caused by nitrates from agricultural sources. *Official journal of the European Communities*, **L375**: 1–8 (1991).
220. Council directive 76/464/EEC of 4 May 1976 on pollution caused by certain dangerous substances discharged into the aquatic environment of the Community. *Official journal of the European Communities*, **L129**: 23–29 (1976).
221. Council directive 80/68/EEC of 17 December 1979 on the protection of groundwater against pollution caused by certain dangerous substances. *Official journal of the European Communities*, **L020**: 43–48 (1980).
222. Council directive 75/440/EEC of 16 June 1975 concerning the quality required of surface water intended for the abstraction of drinking water in the Member States. *Official journal of the European Communities*, **L194**: 26–31 (1975).
223. *Alma-Ata 1978*: *primary health care*. Geneva, World Health Organization, 1978 ("Health for All" Series, No. 1).
224. *Targets for health for all. Targets in support of the European regional strategy for health for all*. Copenhagen, WHO Regional Office for Europe, 1985 (European Health for All Series, No. 1).

225. *Health for all targets. The health policy for Europe.* Copenhagen, WHO Regional Office for Europe, 1993, p. 228 (European Health for All Series, No. 4).
226. Council directive 96/61/EC of 24 September 1996 concerning integrated pollution prevention and control. *Official journal of the European Communities*, **L257**: 26–40 (1996).